The
REAL BIBLE
and
SCIENCE

RICHARD A. LANGEN

ISBN 978-1-64299-844-3 (paperback)
ISBN 978-1-64299-845-0 (digital)

Copyright © 2019 by Richard A. Langen

All rights reserved. No part of this publication may be reproduced, distributed, or transmitted in any form or by any means, including photocopying, recording, or other electronic or mechanical methods without the prior written permission of the publisher. For permission requests, solicit the publisher via the address below.

Christian Faith Publishing, Inc.
832 Park Avenue
Meadville, PA 16335
www.christianfaithpublishing.com

Printed in the United States of America

This book uses quotations from the NIV (New International Version) of the bible. Copyright requirements for quotations conform to the requirements of the originator as follows:

THE HOLY BIBLE, NEW INTERNATIONAL VERSION® NIV®
Copyright © 1973, 1978, 1984 by International Bible Society®
Used by permission. All rights reserved worldwide.

The text may be quoted in any form (written, visual, electronic or audio), up to and inclusive of five hundred (500) verses without the express written permission of the publisher, providing the verses quoted do not amount to a complete book of the Bible nor do the verses quoted account for 25 percent (25%) or more of the total text of the work in which they are quoted.[1]

---

[1] https://www.studylight.org/site-info/copyright/bible/en_niv.html

Contents

Preface 9
    The Mission of This Book 9
    Staying Humble 11

Introduction 13
    The True Bible 13
    Key Erroneous Contemporary Beliefs 14
    Science Agrees with the Timeline of Genesis 15
    Days 1–4 of God's Work: Correcting
    Conditions Affecting the Earth 15
    Civilizations and Competition Increase: Adam
    Made King 16
    The Flood and Reboot of the Earth 16
    Living Things and Evolution 17
    The Formation of the Universe 17

1   Modern Bible Errors and Concepts 19
    What is the Bible? 20
    The Nature of the LORD and God 21
    Creation of the Universe Not Part of the First
    Verse of the Book of Genesis 24
    The Earth "Became" Empty and Void 26
    Contemporary Beliefs: A Fertile Source of False
    Evidence for Critics 26
    No Creation during God's Six Days of Work 27
    Adam's Physical and Spiritual Formation Dates 28

	A Major Inconsistency Disappears 29
	How Genesis Fits into the Formation of the Universe 29
	Belief without Faith 32
2	The Genesis Timeline and Prior Events........................33
	The Gregorian and Hebrew Calendars 34
	Ice Core Studies: A Look into the Past 35
	Biblical Events Compared to Events on Earth 39
	Disasters Leading up to God's Work 41
	Recent Events on Earth..44
	Science Proves the Bible 47
3	Days 1–4: Earth Recovers 48
	Comet Strikes—A Key Factor 48
	Day 1: Light Appears 51
	Earth's Atmosphere and Magnetic Field 54
	Day 2: The Separation of Waters..............................57
	Day 3: The Sky Clears and the Land Dries................59
	Day 3: Plants Recover 60
	Day 4: The Mist Clears and the Sun Comes Out 62
	The Bible and Science Agree 63
4	The Start of Civilization..64
	Earliest Man Preexisted God's Six Days of Work 65
	Earliest Settlements Show Advancing Bad Conditions....67
	God Intervenes to Improve Man's Existence................70
	God Solves the World's First Labour Problem 73
	Formation from Clay Does Not Debunk Evolution 75
	Animals and Adam's Helper 82
	Eve, Fall from Grace, Adam Starts a Family................84
	Cain Leaves and Interacts with Existing Humans on Earth 85
	Science Enhances the Bible 88

5	Civilization Grows: Adam Becomes King	89
	Cain: Rogue or Nation Builder?	90
	The Moving Persian Gulf and Crowding	91
	Day 5: God's Response to Increasing Needs	93
	Day 6: More Resources for a Growing Population	94
	Day 6: Handing over Dominion	96
	The Uruk Period and the Sumerian Culture	99
	The Sumerian King List: Adam the First King	103
	Kingdoms Change: Resistance toward the LORD	105
	The Final Defilement	107
	Society Becomes More Corrupt	109
6	The Flood and Restart of Life	110
	Flood Preparations: The Ark Passenger List Changes	111
	The Flood Comes	113
	Flood Story Detractors and Supporters	115
	Confirming the Science of the Deep	118
	After the Flood: A Long Drying Process	122
	Answers on the 168 Lost Years	127
	Noah's Sons and Grandchildren: The First Kings after the Flood	127
	The Temple of E-ana at Uruk: The Tower of Babel	129
	The Recurring Competition among Men	130
	Abraham: The New Root of the LORD's People	131
	Semitic Laws, Institutions, and Traditions: Lasting Legacies	133
7	Living Things and Evolution	135
	The Biblical Jump in Time	135
	DNA and RNA and Evolution	138
	The First Life on Earth	140
	More Complex Organisms as the Earth Changes	142
	Early Animals	144

	First Vertebrae Animals	146
	Transition to Land	147
	The Age of the Dinosaurs	151
	Mammals: The Door for Manlike Animals	152
	The Primate Order – Where Man Came From	155
	Man Emerges	158
	Homo sapiens	162
8	Formation of the Universe and Earth	170
	The Big Bang: Physical and Spiritual Formation	170
	The Timelessness and Infinite Size of the Universe	175
	The Book of Proverbs	176
	The Solar System and Earth Appear	177
	Formation of the Deep	180
	Land and Then Oceans	184
	Continents Form – Oceans Deepen	187
	Formation of Soil	190
	How Does the Bible Fit So Well with Science?	191
	The LORD's Master Plan	193
9	Conclusion	197
	The Bible Is the Living Word	197
	For Those Who Want to Explore More	198
Citations		201
Index		203

Preface

The Mission of This Book

Figure P-1 – My Family in Their Sunday Best and Ready for Church

My religious upbringing was Roman Catholic, but my situation is probably like most, if not all, those brought up in contemporary religions today. My family, by local standards, followed the faith closely. When I was a young child, we accepted the teachings of the church as gospel. There was no room to question the church. Rationalization about why the church was right on any issue was possible but not encouraged. Questioning things came with risks. The safest way involved minimal questions and the acceptance of teachings as verbatim.

During my childhood and adolescence, formal religious instruction generally centered on the teachings of the church rather than the Bible. Our religion teachers were more likely to use the latest papal encyclical rather than the Bible. We had limited knowledge of the Bible. Catholic dogma and rituals, rather than the Bible itself, colored our religious education.

Postsecondary education, experiences in life and other religions influenced my attitude toward contemporary religions. My grandfather probably considered a graduate from grade 8 in a single-room country school as highly educated. In those days, education was not as important as it is today. Work on the farm guiding a team of horses did not require too much education.

During my generation, things changed. People began to question the teaching of contemporary religion. They refused to fall into lockstep with the all-knowing and unyielding clergy of old. Many of my relatives and friends, including my parents, continued following the church, but many others did not. This seems reflective of society in general. Some follow contemporary religions wholeheartedly. Some seemingly stay in religions as a matter of family tradition but grudgingly follow under protest. Some have fallen away from the church. Others vehemently object to church teachings but still believe in Christianity. Still, others are agnostics and are not even sure if God exists; and some are atheists, who believe that there is no God.

I have talked to others, including some of my children, who are skeptical of religion. In many cases, they are also skeptical of the Bible and Christianity in general. The common theme is that religions are out of touch with reality. They often cite religious teachings that fly in the face of scientific knowledge. Generally, they confuse the teachings of contemporary religion with the Bible itself. Their questions about religious teaching lead them to throw the baby out with the bathwater. They trash the Bible along with religion.

Generally, most Christian religions acknowledge and follow the Bible. If we start with the contention that the Bible is correct, then divergent teachings among religions proves that some or all religions are on the wrong track. Contradictions in teaching among religions only fuel the skepticism among those who question religion. We need to ignore the teachings of religions for a while and concentrate on the Bible itself. If we can establish the Bible as correct, then we have a hope of reestablishing a belief in God and perhaps in religions as well.

What accounts for this general skepticism and questioning of contemporary religions? Is the skepticism valid? I believe that the skepticism about religion is valid. Certain aspects of my early religious education "fly in the face" of every known scientific fact available. I cast no dispersion on the Catholic nuns who taught me my first formal religion classes. They were loving individuals who would do anything for us kids. They had no choice but to follow the teachings of the church in lockstep.

I wrote this book to help reestablish the credibility of the Bible. I believe in both science and the Bible. If I can show that the Bible agrees with science, perhaps I can play some small role in bringing back people, such as some of my children, to the LORD God. The first step in this process is to reestablish the Bible as valid.

Staying Humble

While I have some problems with the teachings of religions, I by no means condemn or disdain them. I hope my book does not appear to pontificate about high and mighty things. I still have many friends and relatives who still follow religion closely. I have the utmost respect for them, even though I no longer subscribe to any one religion. I do though share a profound belief with them in Christianity and the LORD God. I am subject to the same human weakness of pride as anyone else. This book contains the word "I" when referring to the work carried out. It is tempting for me to think that I came up with all the subject matter of this book. This would give me much more credit than I deserve.

This book was a long, learning journey for me. I started out with some questions and began to study and write. Many learnings did not come easy. One thought, after some delay, led to another thought. I am sure that these thoughts did not come from me. Some other guiding force was leading this lowly, undeserving individual along to the next thought. I give credit for this to the LORD God.

Introduction

Many believe that the Bible is out of touch with reality and that the Bible runs contrary to modern science. I do not believe this. My study of the Bible as outlined in this book proves the Bible, as originally written, is entirely in agreement with modern science. The modern translations of the original Hebrew Bible texts are in error in many cases. Many centuries of erroneous religious dogma and doctrine have exacerbated the problem. As a result, many people today hold the Bible in disrepute.

The True Bible

In writing this book, I looked at the original Hebrew Bible. I found that some parts of modern translations of the Bible do not represent what the original Hebrew text says. This, in part, is a huge contributing factor that leads to discrepancies and contradictions between what modern Bibles say and what science says. Throughout this book, you will find numerous references to what modern Bibles say and what the original Hebrew biblical text says. Anyone interested in seeing the original Hebrew text can do so by going to biblehub.com.

I make numerous references to Hebrew words. I obtained the definition of biblical Hebrew words from concordances. Concordances provide a detail definition of each word. Concordances base the word definition upon the Hebrew language, as spoken by Hebrews. The difference between the actual meaning of some Hebrew words and what we find in modern Bible translations is truly amazing. You don't have to take my word for anything. You can look up the meaning of these at biblehub.com or other similar websites. I generally used

Strong's Exhaustive Concordance (Strong n.d.) to determine the meaning of Hebrew words in the original Bible. Bible Hub provides direct access to many concordances.

Key Erroneous Contemporary Beliefs

Chapter 1 looks at some important erroneous beliefs that result from faulty translation of the original Hebrew Bible. The following summarizes some of these faulty beliefs.

- The Nature of God – Most contemporary religions refer to the LORD and God as the same. This is not valid.
- The First Verse of the Bible Includes No Creation – Many creationists rely on the erroneous translations of modern Bibles as a basis that God created all things from nothing. Contrary to popular belief, the first verse of chapter 1 of the book of Genesis does not mention the "creation" of the universe or anything else.
- The Earth Became Formless and Empty – The footnotes in some modern Bibles elude to the correct meaning of the second verse of the book of Genesis. The Earth "became" formless and empty rather than "was" formless and empty, as stated in most modern translations of the Bible.
- No Creation During God's Six Days of Work – My religious education taught me that God created all plants on day 3, all fish and birds on day 5, and all animals on day 6. We will see that the original wording of the Bible does not support this view.
- Adam's Physical and Spiritual Formation Dates – Many Christians believe that God created Adam on day 6. Not true; God formed Adam from the dust of the Earth on day 3. On day 6, God imparted a resemblance of his spirit to Adam.
- How Genesis Fits into the Formation of the Universe – Many believe that the book of Genesis is a detailed narrative of the universe, starting with its creation. Not true;

the book of Genesis covers mainly the last fifteen thousand years or so of the Earth's existence.
- Chapter 1 of Genesis Interconnects with Chapters 2, 3, and 4 – Skeptics believe that different authors wrote two different versions of the book of Genesis, one in chapter 1 and different versions of the same narrative in subsequent chapters. Not true; chapter 1 runs in parallel with subsequent chapters.

Science Agrees with the Timeline of Genesis

Chapter 2 shows how the narrative of the Bible matches Earth events as described by modern science. We get a new knowledge of what the term "formless and empty" in the book of Genesis means. I show how science definitively proves that the book of Genesis covers mainly the last fifteen thousand years or so of the Earth's existence.

Days 1–4 of God's Work: Correcting Conditions Affecting the Earth

Unlike the teachings of many religions, days 1–4 of God's work did not involve the creation of the Earth, sun, moon, and stars. In chapter 3, we see how these days were days of healing, when the Earth recovered from the events that made it *become* "formless and empty." The Earth, sun, moon, and stars preexisted the detailed narrative of the book of Genesis by billions of years.

In chapter 4, we see that Adam was not the first man on Earth. Science tells us of Homo sapiens that existed on Earth, long before the six days of God's work.[1] We will look at Mesopotamian history, and we will see that the Bible intimately intertwines with the history of Mesopotamia, the "cradle of civilization." The arrival of Adam and his descendants coincides with a remarkable increase in the civ-

---

[1] C. Beard, *The Hunt for the Dawn Monkey* (Berkeley and Los Angeles, California, USA: University of California Press, 2004).

ilization and development of Mesopotamia. It was no accident that Adam's arrival ushered in this era.

Chapter 4 shows us the location of the Garden of Eden. We will see that teaching of contemporary religions includes a nearly three-thousand-year error for the formation date of Adam. As well, a look at science proves that the Garden of Eden could not have possibly existed on day 6 of God's work. God expelled Adam and Eve three thousand years earlier than day 6.

Civilizations and Competition Increase: Adam Made King

Contemporary religions teach that God created fish and birds on day 5 and animals on day 6. Chapter 5 will show that this is not the case. No creation occurred, and God merely decreed that certain life that serves as food for man increase. This was to satisfy the increasing needs of society as the great societal advances of Adam occurred. We will see that Mesopotamian history reflects how the LORD gave Adam dominion over the world, including its people. Chapter 5 discusses how the weaknesses of human nature got the best of man, leading the LORD to wipe out mankind and start over.

The Flood and Reboot of the Earth

Chapter 6 discusses the coming of the flood and the destruction of nearly all life on Earth. We will see how science provides evidence of the great flood and how the flood took hundreds of years to fully recede from Earth. Mesopotamian history again coincides with the Bible, as human life reestablishes itself on Earth. Noah's descendants became kings and resumed kingship of the world. We will see how stories in ancient Mesopotamian tablets mirror biblical stories, such as the Tower of Babel. The LORD God confused the speech of inhabitants, causing them to separate and scatter. Many formed their own countries and states.

Living Things and Evolution

Chapter 7 looks at evolution. We show how evolution undoubtedly is a valid theory and is not in conflict with the Bible. We see how fast the world advanced with the LORD's direct intervention through Adam. Nature needed billions of years to develop life on Earth, to the primitive Homo sapiens prior to Adam's arrival. With the LORD God's help, man needed only a few thousand years to evolve to the modern world we see today.

The Formation of the Universe

While most of the Bible tells us about life on Earth starting about fifteen thousand years ago, some verses do tell us about the very beginning of the universe, starting with the Big Bang. The Bible alludes to the early processes that formed the universe and Earth. This proves the Bible is inspired by the LORD as no science existed to guide biblical writers on these matters. Chapter 9 looks at the many apparently random processes that brought about the Earth. It is difficult to imagine that the Earth, a marvel of nature that supplies all our needs, was a purely random process. The LORD's six-billion-year plan culminated in the Earth we enjoy today.

1

Modern Bible Errors and Concepts

Many believe the Bible is a totally irrelevant book and have discarded it as a source of truth. By writing this book, I came to find that the Bible is not irrelevant. Much of the skepticism involves supposed contradictions between science and the Bible. I found this belief is totally unfounded, as I will discuss in subsequent chapters of this book.

Part of the skepticism results from modern translations of the original Hebrew Bible. Some scholars conclude that the Bible contains many errors and contradictions. Even the Roman Catholic church apparently has misgivings about parts of the Bible. Dr. Maurice Bucaille (Bucaille n.d.), in his book *The Bible The Qur'an and Science*, discusses the words of His Grace Webber in finalizing the texts of the second Vatican Council. (Bucaille n.d.) His Grace Weber feels that the book of Genesis contains material that is imperfect and obsolete. In describing parts of the Bible as "imperfect and obsolete," it appears that even the Catholic church has discarded the beliefs of their own St. Augustine—that the Bible is infallible. A major reason for turning away from the Bible lies in the description in the book of Genesis of the creation, the six days of God's work, the great flood, and various other subjects.

According to an Internet article[1] by Rev. Nicanor Pier Giorgio Austriaco, OP, even the present pope, then Cardinal Ratzinger, has

---

[1] http://www. hprweb. com/2009/01/reading-genesis-with-cardinal-ratzinger/

problems with the book of Genesis. The article states that Cardinal Ratzinger feels that the explanation of Genesis regarding how the world arose does not and cannot provide a scientific explanation of events. According to Cardinal Ratzinger, the Bible seeks to describe profound religious realities rather than accurate scientific fact. Cardinal Ratzinger's advice includes reading Genesis not from a scientific point of view but from a religious point of view, discussing profound truths about the Creator. The article states that some disagreement exists within the Roman Catholic church regarding this issue.

I disagree with Cardinal Ratzinger. I believe the Bible is the inspired and irrefutable word of the LORD God. I find that the modern translations of the original Hebrew Bible are responsible, in great part, for many supposed contradictions and errors. When we examine what the original Hebrew biblical authors wrote, a new picture of the Bible emerges. The contradictions and errors disappear. This chapter discusses some of the basic errors in contemporary beliefs that contribute to the supposed errors and contradictions.

What is the Bible?

Before I discuss errors and concepts, a brief description of the Bible is in order.

Religions and faith systems generally record their beliefs in special books or texts, sometimes called holy books. Most religions, such as the Buddhists, the Hindus, and the Bahá'í faith have holy books that have no relationship with Christian religions. Muslims acknowledge and believe in some aspects of the Bible but follow their own holy book, the Quran.[1] The holy book of Judaism or the Jewish religion is the Tanakh.[2] Jews refer to the first five books of the Tanakh as the Torah.[3] The Torah contains references to the universe, Earth, and the establishment of life on Earth.

---

[1] https://en. wikipedia. org/wiki/Islamic_holy_books
[2] https://en. wikipedia. org/wiki/Tanakh
[3] https://en. wikipedia. org/wiki/Torah

The Bible is the universal holy book of Christians across the world. The Bible consists of two main divisions, the Old Testament and the New Testament. The Old Testament is the Jewish Tanakh. Scribes wrote the New Testament less than two thousand years ago, which mainly describes the life and teachings of Jesus Christ. The Bible consists of over sixty-five books or major sections. My work refers mainly to three books of the Bible, namely Proverbs, Psalms, and Genesis.

The ancient biblical figure King Solomon wrote most of the book of Proverbs. History describes King Solomon as very wise. Purportedly, he asked God for wisdom, and God gave him wisdom. King David of Israel wrote the book of Psalms. Psalms and Proverbs mention aspects of the formation of the universe and some of the major phases of the Earth's geologic development, discussed in detail in chapter 8. We do not know who authored Genesis. Genesis mainly covers roughly the last 15,000 to 4,500 years of the Earth's history. Over the years, many translations of the original Hebrew texts emerged into the modern Bibles Christians follow today.

The Nature of the LORD and God

The Bible uses the terms LORD and God. Mark Harris (Harris n.d.) discusses this in his book *The Nature of Creation*. Chapter 1 of Genesis uses the term "God" exclusively, while other parts of Genesis use the word "LORD." Scholars assert that different authors account for this change in names and that each author presented a separate and different story. This assertion, among a host of other assertions, leads them to conclude that parts of Genesis are a mottled collection of assorted writings that do not relate to each other. This is a major reason why many protagonists write off much of Genesis as a futile collection of unconnected verses that have no foundation or common source. Harris refers to the authors of the first chapter of Genesis, which refers to the name God, as the "priestly" story. He refers to other parts of the Bible that refer to YAHWEH as the "Yahwist" story.

The belief that the 'LORD' and 'God' are the same being is a major error. The words 'LORD' and 'God' are basically different. Recognition of this error eliminates many of the supposed contradictions scholars see in the Bible. Also, we will see that this belief leads to a collection of other mistaken beliefs.

Some parts of the Bible use the name LORD God. Just as in human organizations, various personnel can work independent of with their supervisors. So too in the Bible. A president and a vice president attending a board of directors meeting is synonymous to actions in the Bible involving the LORD God. Just as the president and vice president act together, so does God act together with the LORD. But the vice president is a separate person from the president and may attend other meetings and make decisions without the president. This is synonymous to parts of the Bible describing actions of God only, as in chapter 1 of Genesis. But just as in human organizations, the president can act alone or overrule the vice president. So too in the Bible, the LORD acts alone. This occurred before the flood, when the LORD intervened to change the numbers of certain animals entering the ark. We will see below that human organizations differ from the LORD and God. Unlike human organizations, God flowed forth from the LORD, when the LORD brought about the universe.

The book of Proverbs tells us about a time before the Big Bang, when the LORD brought about the universe. The relevant verses open with the introduction of Wisdom, a spiritual entity, brought forth before the big bang: "The LORD brought me forth as the first of his works, before his deeds of old. I was formed long ages ago, at the very beginning, when the world came to be."[1] Wisdom came about before all other things, including the universe. Before Wisdom, there was only YAHWEH or the LORD, as translated in modern Bibles.

The original Hebrew wording of the Bible for "long ages ago" alludes to the earliest possible time, or perhaps a time so early that time vanishes. The word alludes to a point where perhaps time was

---

[1] Proverbs 8:22–23

meaningless and nonexistent. The Bible talks about a very long time before the Earth formed and even before the universe formed.

The original Hebrew text for "brought me forth" really means "owned me." Thus, it seems that Wisdom was not a created entity, but rather a quality the LORD possessed or owned. The LORD brought Wisdom forth from himself or poured Wisdom out as one of his first works. Thus, Wisdom is the very first of the spiritual beings emanating from the LORD. Before Wisdom, there was nothing; only the LORD.

In chapter 8, I will discuss how Proverbs goes on to list things not in existence during the time covered by the verses above. We will see that the time was before the formation of the heavens and universe.

The book of Psalms also mentions the same period, discussed above in the book of Proverbs. "By the word of the Lord the heavens were made, their starry host by the breath of his mouth."[1] I discuss this verse again in chapter 8, but for now, I will concentrate on the last part of the verse: "their starry host by the breath of his mouth."

Modern Bibles totally change the original Bible text with the inclusion of the word "starry." The original Hebrew Bible does not include the word "starry." Modern Bible interpretations lead us to believe that these verses talk about the formation of stars, but this is not the case. The original Hebrew Bible verses correctly translated should read "by the breath of his mouth, all the host." The original Bible uses a Hebrew word for "host" that means a legion or army, with the connotation of a confrontation or a campaign. I call this large army of beings the "heavenly host." Along with the universe, the LORD brought forth a "heavenly host" of spiritual beings.

The LORD or YAHWEH is the singular, never-ending, timeless being that exists solely of his own volition. YAHWEH or the LORD is the omnipotent all-powerful being that initiated the big bang. As indicated, YAHWEH also brought forth a multitude of spiritual beings. Some of these beings were undoubtedly angels, but some of them were beings that we call God. As the Bible tells us above, before the Big Bang, there was only the LORD. God came about as part of the Big Bang. We will see in chapter 8 that the

---

[1] Psalms 33:6–7

universe and the heavenly host are the LORD and the LORD is the universe and the heavenly host. The LORD did not make anything, but the universe and the heavenly host flowed forth from him.

When we see that the LORD and God, are different, separate beings, the "priestly" and "Yahwist" distinctions discussed previously disappear. This realization helps to remove the argument that the book of Genesis is a disjointed collection of separate stories. As well, the distinction also allows us to decipher the real meaning of other verses in the Bible. As discussed below, the first verse in the Bible is one such verse.

Creation of the Universe Not Part of the First Verse of the Book of Genesis

> In the beginning God created the heavens and the Earth.[1]

Before I wrote this book, I assumed that this verse referred to the creation of the universe and the Earth. Nothing is farther from the truth. A more accurate translation for the first verse in the Bible is "seeking a bountiful place God chose the best of the universe, more specifically fields of the Earth." An example of a similar sentence is "seeking the sweetest food, Dave chose the orchard, more specifically the cherry tree."

As discussed in the previous section, the LORD brought about the universe. The original Hebrew Bible uses the name YAHWEH or יְהוָֹה according to the Hebrew alphabet. The first line of Genesis uses the name "God." The original Hebrew Bible uses the name "Elohim," or אֱלֹהִים according to the Hebrew alphabet for the name "God." The two names are not remotely similar and refer to different beings.

The original Hebrew Bible uses one word for the English words "in the beginning," which means "first bounty" or the first produce or fruit. God noticed that of all places in the universe, the Earth was

---

[1] Genesis 1:1

a place with the conditions to allow a bountiful harvest of plant and animal riches. It had the qualities to result in more than what the rest of the universe was—a rocky and foreboding place.

The Hebrew word for "created" does not mean "created" in the contemporary sense of the word. The Hebrew word used by the original Bible means "to select" or "to decide on." The selection is not just an ordinary selection, but a process of picking the best of the best or the best of the good. The word has a connotation of making prosperous or affluent, in the sense of well-fed or well-nourished.

The Hebrew language has no word for the English word "and." In the scripture shown above, the original Hebrew authors placed a word between the words "heaven" and "Earth," which means "more specifically."

The selection process probably resembles how many farmers chose a location for their farmyards in my home province of Saskatchewan. The early settlers cleared the land of trees to make way for crops. However, they did not clear all the trees. They picked locations with tall trees for shelter, well-drained to keep things dry and land with the best soil for lush gardens. Many farmyards had fruit trees such as plum trees and crab-apple trees and other fruit. I remember my mother. She was very good at gardening and was famous throughout the relationship for her chicken dinners. All the fresh vegetables for these dinners came from her garden. After dinner, she toured people through the garden, which had colorful flowerbeds as well as vegetables. I can remember playing in a treed area beside the garden, where many berry trees provided delicious snacks. Yes, those were the good old days. We ate like kings and relaxed close to nature. Those old settlers certainly did a good job in picking out a gem from their land to live in. As I think back, these places were mini Gardens of Eden. God used the same process when he picked the Earth from all other planets in the universe.

The Earth "Became" Empty and Void

> Now the Earth was formless and empty, darkness was over the surface of the deep, and the Spirit of God was hovering over the waters.[1]

This translation leads one to the conclusion that this verse talks about the initial formation of the Earth. This is not the case. This verse talks about a time much later.

A footnote at the bottom of my NIV Bible is one of the things that started me thinking about how the book of Genesis fits into the history of the Earth. The second verse in Genesis states that the Earth "was" formless and empty. The footnote states that another translation of the word "was" is "became." A study of the original Hebrew Bible confirms that "became" is the correct interpretation. The Earth "came to be" or "turned into" the formless state in the Bible. Only a preexisting Earth could "become" formless and empty. We will see in chapter 3 that a huge calamity befell the Earth, causing it to become formless and empty.

The original text of the Bible tells us that the calamity that befell the Earth tells us that the Earth became a place barren, devastated, spoiled, and of no worth. Disorder and chaos were the order of the day. As the Hebrew word for "darkness" implies, the earth became a place of gloom, wretchedness, despair, suffering, and devastation; much more than just the absence of light. The wording of this verse provides an inkling that life already existed on Earth. Gloom, wretchedness, and despair can only apply to living things. As we will see in chapter 4, science tells us that humans existed during this time.

Contemporary Beliefs: A Fertile Source of False Evidence for Critics

Creationists believe that the Bible tells us that God created everything from nothing. This leads to many supposed biblical errors and contradictions. Creationists also generally adhere to the belief

---

[1] Genesis 1:2

that the first verses of Genesis describe the creation of the universe and the Earth. This use of a faulty translation leads to many unanswerable questions. Creationists have a great deal of trouble justifying the appearance of the sun, moon, and stars on day 4 of God's work. According to creationists' literal interpretation of modern biblical translations, God created the Earth first and then the sun, moon, and stars later. This leads to one hard question: Where did the light come from on day 1 of God's work? These inconsistencies provide fertile ground for critics.

A myriad of other errors and inconsistencies arise from contemporary beliefs about the Bible. Doctor Maurice Bucaille[1] and Mark Harris[2] discuss these inconsistencies and error in detail. I do not want to get into the details of these errors and inconsistencies here. I recommend that anyone interested in the details of these errors and inconsistencies read their books.

Mark Harris sums up the resulting confusion quite well in his book. He refers to Barton and Wilkinson who feel that using the early chapters of Genesis literally is naive at best and in a simplistic way totally ignore scientific questions. They feel that many Christian authorities have even lost confidence in Genesis. Harris feels that most Christians steer clear of meaningful discussion and resort to a more lyrical and figurative interpretation of Genesis, totally ignoring tangible science. Subsequent chapters will show how these criticisms and contradictions disappear in the light of the true meaning of what original Bible authors wrote.

No Creation during God's Six Days of Work

As discussed in appendix 1, some dictionaries include a definition for the word "create" that I call "magic wand theory." Mark Harris characterizes modern religion's view of creation as *ex nihilo*, where God magically created the world from nothing and that God is not dependent on the world.

---

[1] D. M. Bucaille (n. d.). *The Bible The Qur'an and Science.*
[2] M. Harris (n. d.). *The Nature of Creation.*

In chapter 5 I will show that the Bible uses no words that fit the "magic wand" definition in some dictionaries. Chapter 1 of Genesis tells us that God decreed that agricultural plants increase copiously on day 3 of his work. On days 5 and 6 of his work, God decreed that certain types of fish, birds, and animals suitable as food for man increase bountifully. Science tells us of climatic conditions of the time that restricted available food supplies for an ever-expanding population. God responded with a decree to increase man's food supply.

Adam's Physical and Spiritual Formation Dates

The physical formation of Adam happened on day 3 of God's work, approximately three thousand years before God made Adam and Eve in his own images. In chapter 3, I will discuss in detail how plants started to grow again on day 3 of God's work. We will see how seeds fell on the surface of the ground. The narrative in chapter 1 of Genesis says the seeds sprouted and grew. Chapter 2 of Genesis flashes back to chapter 1, to a time when "no plant had yet sprung up." Chapter 2 of Genesis clearly points to a time after the seeds fell on the ground but had not yet sprouted. The LORD needed someone to cultivate the ground between the time seeds fell on the ground and when the seeds sprouted. He formed the physical Adam, or man of low degree, from the dust of the ground to fulfill the need for someone to cultivate the ground. Chapter 1 of Genesis states that plants came up before the end of day 3. Thus, Adam's formation date was in day 3. Ordinarily, seeds on the ground would sprout if rain fell. Chapter 2 of Genesis tells of special conditions why seeds would not sprout and why God needed Adam to cultivate the ground.

Chapter 5 of Genesis gives an account of the genealogy of Adam and his descendants. The chapter opens by defining the starting point for the genealogy as the time when Adam and Eve received God's spirit.

A Major Inconsistency Disappears

The previous section shows how chapter 1 of Genesis intertwines with chapter 2 of Genesis. This along with the fact that the LORD and God are different beings solves a major criticism of detractors of Genesis.

Doctor Maurice Bucaille (Bucaille n.d.) is typical of critics who rely on contemporary translations of the Bible. In his book, he includes two headings in his book, discussing what he calls the first and the second descriptions of creation. According to him, the Bible provides two different versions of creation, one in chapter 1 of Genesis and another in chapter 2.

Chapter 1 uses the name "God" only, while chapter 2 uses the name "LORD." According to Dr. Bucaille, this is an inconsistency. He feels that different authors used different names for what contemporary religions commonly call "God." This criticism disappears when we realize that the LORD and God are different beings.

Dr. Bucaille cites chapter one of Genesis as masterfully inaccurate from a scientific point of view. Subsequent chapters will demonstrate that this is not the case. Dr. Bucaille's view that chapter 1 conflicts with chapter 2 of Genesis makes abundant sense, if we assume that each chapter is a different story about creation. However, the underlying assumption that the Bible presents two descriptions of creation is false. As discussed in the previous section, chapter 1 of Genesis presents a summary of the six days of work. Chapter 2 fills in some of the details of what happened during the various steps of the work.

How Genesis Fits into the Formation of the Universe

Chapter 2 of Genesis starts with a short description of how the detail events of Genesis fit into the overall timeline of the formation of the universe.

> This is the account of the heavens and the Earth when they were created, when the LORD God made the Earth and the heavens.[1]

---

[1] Genesis 2:4

At first the first part of the verse seems repetitious with the second part, but the Hebrew words used explain that it is not. The Hebrew word used for "account" means ancestry or generations. The universe went through many stages or generations of formation in the fourteen billion years between the Big Bang and the formation of the Earth we know today. The original Hebrew wording of the Bible tells us that the first word "heavens" means the totality of space including the moon, sun, and stars or the universe. The first word "Earth" means the Earth with its productive land, as we see it today. As discussed previously the Hebrew word for "created" means "to select" or "to choose." Thus, the first part of the verse tells us that this is a story about generations of the universe, starting with the formation of the universe many billions of years ago and ending with the Earth as we know it today, billions of years after the Big Bang.

The next part of the verse, "when the LORD God made the Earth and the heavens," specifies the exact time or part of the generations the Bible covers. The Hebrew word used for the second word "heavens" means heavens closer to Earth. The Hebrew word used for the second word "Earth" means the Earth in geographical sense, with the continents as we see them today. This part of the verse limits the story to only the recent history of the Earth, which includes a time after the continents formed, as we see them today.

Figure 1-1 shows how the Earth went through many super continent phases. The continents of the Earth moved around the globe, to arrive at the locations we see them at today. The Bible covers a time when the continents resembled the last illustration, captioned "Present Day" in figure 1-1 (Springer 2007). Present-day continents move much less than in previous generations of the Earth's development.[1]

---

[1] Springer. *The Black Sea Flood Question:*. Dordrecht, The Netherlands, The Netherlands: Springer, 2007.

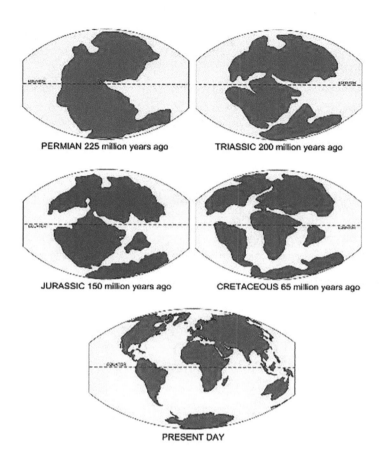

Figure 1-1-Continental Formation of the Earth

The dinosaurs lived about 200 million years ago, during a time captioned as "Triassic" in figure 1-1. The Bible does not mention the time of dinosaurs and so most of Genesis obviously starts after this time. Approximately 60 million years ago, the predecessors of the primates appeared close to the end of the period captioned "Cretaceous" in figure 1-1. Humans formed millions of years after this time. Genesis describes Adam naming the animals of the Earth. Thus, Genesis starts sometime closer to present time, when humans were on the Earth and when the continents move to where we see them today, captioned "Present Day" in figure 1-1. This verse does not provide the exact start date for chapter 2 of Genesis. As we will see in chapter 2, the time covered likely starts less than twenty thousand years ago.

Over the centuries, many erroneous conclusions about the so-called creation in Genesis occurred. In the mid-seventeenth century, James Ussher, archbishop of Ireland, calculated the date of the formation of the universe as the evening before Sunday, October 23, 4004 B.C. In the 1700s, Geologist James Hutton started to change thinking of past times. His *Theory of the Earth*, published in the Proceedings of the Royal Society indicated that humans occurred on the on the planet recently, compared to the multibillion-year history of the universe and long after the appearance of fish, amphibians, reptiles, plants, birds, and mammals.[1]

Belief without Faith

Some proponents, including some clergy, imply that belief in the first chapters of Genesis are more a matter of faith than logic. I strongly disagree. The Bible is completely correct, as originally written. Modern translations of the Bible account for much of supposed inconsistencies in the Bible. This leads to religious dogma and doctrines that farther cloud the picture.

We require no faith to conclude that the Bible is correct and that it agrees with science. We shall see from analysis of science and proper interpretation of the Bible that it stands on its own by fact. We require no leap of faith to believe that the Bible stands correct on its own.

---

[1] W. A. Ryan. *Noah's Flood*. New York, NY, USA: Touchstone, 2000.

2

The Genesis Timeline and Prior Events

A comparison of biblical events to events that shaped the Earth helps us to see that the Bible matches science. Antarctica ice core analysis helps us in this regard by providing a picture of the Earth many thousands of years ago. Comparison of past Earth events to biblical events helps to see what conditions affected the story of the Bible.

Placing biblical events in time is key to interpreting the Bible. Early authors of the original Hebrew Bible went by the Hebrew calendar,[1] which is different from the Gregorian calendar we use today. We must reconcile the Hebrew calendar to the Gregorian calendar. This allows comparison of the timeline of biblical events to the timeline of scientific events.

The comparison reveals amazing results. We see that Genesis mainly covers a time starting approximately twelve thousand to thirteen thousand years ago. We even see the great flood of Noah in the scientific data. A study of Earth events preceding the time of Genesis tells a story of great devastation upon the Earth. We get a new and more enlightened view of what "formless and empty" means in the second verse of Genesis.

---

[1] https://en. wikipedia. org/wiki/Hebrew_calendar#History

The Gregorian and Hebrew Calendars

Unlike the Gregorian calendar, the Hebrew year number reflects the number of years since the so-called creation, starting at the end of day 7. The Bible says God rested on day 7. The year 2000 in the Gregorian calendar is the year 5761 on the Hebrew calendar.[1]

According to Jewish sages, we can understand the history of the world as two cycles, each with seven thousand-year days. The first cycle starts with the beginning of the six days of God's work, as described in Genesis. It ends at the end of day 7 when God rested. The second cycle begins after the seventh day. We are in the sixth day of the second cycle right now.

One important concept is that a day in Genesis equals one thousand years. The apostle Peter tells us about this in the New Testament: "But do not forget this one thing, dear friends: With the Lord a day is like a thousand years, and a thousand years are like a day."[2] As well, Psalms talks about the thousand-year to a day concept: "A thousand years in your sight are like a day that has just gone by, or like a watch in the night."[3]

Gregorian time is linear, while Hebrew time is not. The Gregorian calendar divides the year into roughly 365 days, which recur linearly and indefinitely in time. The Hebrew Calendar uses lunar cycles as a basic unit of measure. Every lunar cycle has 29.5 days. By this measure, a Hebrew year would have 354 days instead of the 365 days, as in the Gregorian calendar. The Hebrew calendar adds an extra month to the year every two or three years to make up for this discrepancy. Hebrews refer to years with added months as leap years, which occur seven times every nineteen years.

We can convert dates on the Gregorian calendar to dates in the Hebrew calendar. This allows us to place biblical events on the geologic time scale. John J. Parsons in his publication *Hebrew for*

---

[1] W. Wolhouse. *The Measures, Weights, and Monies of all Nations*. London, England, UK, 1859.
[2] 2 Peter 3:8
[3] Psalms 90:4

Christians[1] presents a convenient formula to convert Gregorian dates to Hebrew dates. To obtain Hebrew dates to the nearest year, subtract 1,240 from the Gregorian year and then add 5,000—or more simply, just add 3,760 to the Gregorian date.

The timescale used for ice core analysis measures dates backwards from the year 2000. The Gregorian year 2000 is then the year 5760 in the Hebrew calendar. The end of day 7 of God's work on the Gregorian calendar falls 5,760 years before the year 2000. God's seven days of work starts 7,000 years earlier, or 12,760 years ago on the Gregorian calendar. We are now in the Hebrew year 5777 (as of 2017).

There is some uncertainty in this formula, because of the 165 years of captivity of the Jews in Persia. Rabbinical tradition states that these are 165 lost years because the Hebrew calendar possibly excluded them. If the years are indeed lost, then the Gregorian year 2000 becomes the year 5925 in the Hebrew calendar.

Ice Core Studies: A Look into the Past

Scientists have conducted ice core studies at various locations. Most ice core studies took place in Antarctica and Greenland. These locations have the deepest continental ice sheets, remaining from the last (present) ice age. Other ice core studies have also taken place on mountain glaciers such as Mount Logan and in the Arctic.

Ice core studies involve boring into the ice sheet to extract a column of ice, called an ice core. Analysis of the ice core provides information on what occurred in the past. The deeper the sample, the farther back in time the data reflects.

---

[1] http://www. hebrew4christians. com/Holidays/Calendar/calendar. html

Figure 2-1 Team of scientists in Antarctica

Figure 2-2[2] Ice core photographed using fiber optic light

Figure 2-1[1] shows French, Russian, and American scientists in the Antarctica, Vostok team photo holding unprocessed ice cores. The team cut the four to six-meter-long cores coming out of the ice into one-meter sections. The pictured ice columns are unprocessed cores. White containers in the background allow transportation of the one-meter-long sections.

Scientists can date upper portions of ice cores visually, based on color variations in the ice. Visual dating is quite accurate and is possible with ice cores to depths of 1,800 metres or less. The time of Genesis is only twelve thousand to fifteen thousand years ago, a time

---

[1] https://en. wikipedia. org/wiki/File:GISP2_team_photo_core37. jpeg#/media/File:GISP2_team_photo_core37. jpeg– Creative Commons - by NOAA - License – Public Domain

where visual ice core analysis is most accurate. Errors of up to 1,401 years are possible. Charts in future sections show that 1,401 years will not significantly affect any findings in this book. The geologic calendar time scale is so large that 1,401 years hardly shows up.

Figure 2-2[1] shows a nineteen-centimeter-long section of an ice core from a depth of 1,855 meters. A fiber optic source illuminates annual layer structure from below. The section contains eleven annual layers with summer layers (arrowed) sandwiched between darker winter layers.

Many people, including myself until I started researching this book, think the last ice age is behind us. Not so, we are still in the last ice age. As many as five ice ages have occurred in the past. The first was the Huronian Ice Age that occurred over two billion years ago. The last ice age to occur on the Earth is the Quaternary Ice Age,[2] and we are still in it. Ice ages undergo regular cycles of glacial and interglacial activity. Ice advances during glacial periods and recedes during interglacial periods. We are in one such interglacial period right now.

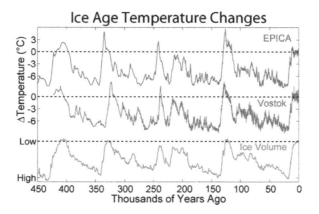

Figure 2-3 Icecore analysis showing temperature anomaly and ice volume

---

[1] https://en. wikipedia. org/wiki/Ice_core– Creative Commons - by NOAA - License – Public Domain
[2] https://en. wikipedia. org/wiki/Quaternary_glaciation

Figure 2-3 shows data from ice core samples obtained from two locations in Antarctica. The green and blue lines show temperature anomalies. A temperature anomaly does not reflect actual temperature, as on a thermometer. The term "temperature anomaly" means a departure from a reference value or long-term average. A positive anomaly indicates that the observed temperature was warmer than the reference value, while a negative anomaly indicates that the observed temperature was cooler than the reference value. This allows a plot of temperature change over time, as shown in the figure 2-3.[1] The top two lines show the estimated temperature change at each of the two locations.

When temperatures rise, ice sheets recede, exposing more land under the ice sheet. When temperatures fall, ice sheets advance, covering up land with ice. Decreasing temperature anomalies with temperatures below zero indicate advancing glacial periods. This leads to cold temperatures and inhospitable weather patterns. The red line shows an estimate of ice volume.

Figure 2-3 shows that over the previous 450 thousand years, we have undergone five glacial/interglacial cycles. Each cycle generally starts with a rapid, fairly steady temperature increase and ends with a rapid, fairly steady temperature decrease. This pattern changed for the last interglacial, as noted in figure 2-4. Something interrupted the temperature rise of the last interglacial, which we are in right now.

---

[1] https://en. wikipedia. org/wiki/Ice_core - Wikimedia Commons - Vostok Ice Cores by NOAA is licensed under Creative CommonsAttribution-Share Alike 3. 0 Unported.

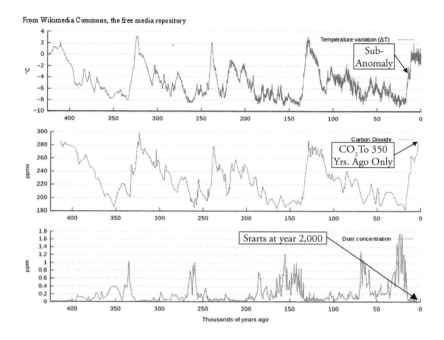

Figure 2-4 Ice core data matches biblical events

Figure 2-4 shows carbon dioxide, dust, and temperature variations for the Vostok ice core over the last 400,000-year period.[1] The notation on the blue temperature curve is very interesting. It shows an interruption in the temperature rise during the present interglacial. I call the temperature interruption a "sub-anomaly." This sub-anomaly seems unique only to the present ice age cycle. It did not occur in previous cycles.

Biblical Events Compared to Events on Earth

Very interesting observations result if we compare biblical events to events on Earth. We find that key biblical events coincide

---

[1] https://commons. wikimedia. org/wiki/File:Vostok_Petit_data. svg- Wikimedia Commons - Vostok Ice Cores by NOAA is licensed under Creative CommonsAttribution-Share Alike 3. 0 Unported. – Note-Time notations made to original illustration

with key events on Earth. Figure 2-5 shows the last twenty thousand years of history.

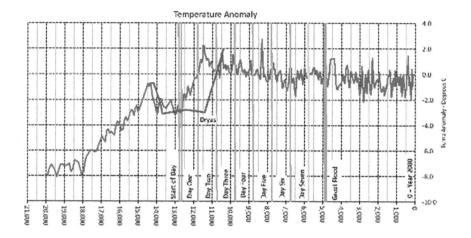

Figure 2-5[1] Events on Earth match biblical events

The vertical lines in figure 2-5 show key biblical events. Pairs of vertical lines show the calculated dates with and without the 165 lost years during the Jewish captivity. The brown line shows a phenomenon known as "Dryas," which I will discuss later.

One very interesting and telling observation is that the calculated start of God's work happens to occur during start of the interglacial that we are in right now. Even more interesting, the start of God's work coincides with the interruption of the temperature rise during the beginning of the present interglacial period. The start of day one of God's work appears to stop the reversal, allowing the previous period of temperature increases to continue. The Great Flood during Noah's time happened approximately 4,900 years ago, using timelines in the Bible. Figure 2-5 shows sharp temperature anomaly changes during this time.

---

[1] Ice Core Data taken from World Data Center for Paleoclimatology, Boulder and NOAA Paleoclimatology Program, AICC2012 800KYr Antarctic Ice Core Chronology, Original_Source_URL: ftp://ftp. ncdc. noaa. gov/pub/data/paleo/icecore/antarctica/aicc2012vostok. txt

It seems like more than just coincidence that God's work coincides with resumption in temperature increases that resulted in the good weather conditions we still enjoy to this day. The one-thousand-year period proposed by the Jewish sages happens to fit exactly with geological events thousands of years ago.

Disasters Leading up to God's Work

A review of the Earth's history reveals many significant catastrophic events leading up to the six days of God's work. The Quaternary extinction event started approximately sixty thousand years before the start of God's six days of work. Between the start of the Quaternary Extinction and the start of God's six days of work, many catastrophic events occurred. A series of comet strikes such as or like the Rio Cuarto Comet strike and other events struck right before the start of the six days of God's work.

Figure 2-6[1] places some of these events on a plot of Antarctic ice core data, showing temperature anomalies at the time. Occurrences generally resulted in temperature decline anomalies. Keep in mind that hundreds and thousands of miles separate Antarctica from the location of the occurrences. This results in delays between the time of the occurrence and the time Antarctica temperatures change. As well, the size of the disturbance created by the occurrence affects the time delay in Antarctic temperature. The Lake Toba Volcanoes[2] had dramatic effects on the Earth. The Toba Lake super volcano erupted in Indonesia.[3] The Toba Volcano probably was responsible for the start of the Quaternary mass extinction.[4] The latest eruption of the

---

[1] Ice Core Data taken from World Data Center for Paleoclimatology, Boulder and NOAA Paleoclimatology Program, AICC2012 800KYr Antarctic Ice Core Chronology, Original_Source_URL: ftp://ftp. ncdc. noaa. gov/pub/data/paleo/icecore/antarctica/aicc2012vostok. txt

[2] http://volcano. oregonstate. edu/vwdocs/volc_images/southeast_asia/indonesia/toba. html

[3] https://en. wikipedia. org/wiki/Toba_catastrophe_theory

[4] C. Chesner. Eruptive history of Earth's large (Toba, Indonesia) clarified. Geology, March 1991.

volcano, referred to as the "Young Toba Tuff," erupted seventy-three thousand years ago. Scientists estimate that the Young Toba Tuff produced an estimated volume 2,500 kilometers³ to of 3,000 kilometers³ of lava that flowed onto the ground around the volcano. The Young Toba Tuff resulted in ash layers as thick as one meter near the volcano and at various locations in Malaysia. A few minor eruptions occurred since the major eruption seventy-seven thousand years ago. The area is seismically active, as evidenced by recent earthquakes. Refilling of the lava chamber of the volcano has uplifted the area, pushing some islands above the elevation of Toba Lake.

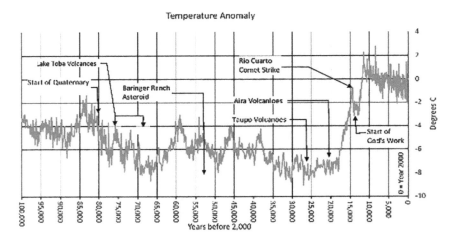

Figure 2-6 Temperature data and catastrophic events before God's six days of work

When Mount St. Helens erupted in the northwestern United States on July 10, 2008, I lived in Calgary over 950 kilometers away. Even at this distance and beyond, the effects of Mount St. Helens were very noticeable, often partially obscuring the sun. Ash from the volcano travelled all the way to Calgary and into adjoining provinces. It is amazing that the volume of the Mount St. Helens volcano was only one cubic kilometer. I can't imagine the effect of the Toba Lake volcano, which was 2,800 times as larger.

Scientists recovered ash from the Toba eruption in the Bay of Bengal in India some 3,200 kilometers away[1]. Later research determined that the volcano deposited an ash layer 150-millimeter-thick all over South Asia. Scientists discovered ash layers as thick as six meters in India and nine meters thick in Malaysia. Scientists originally estimated the height of the eruption column as high as fifty to eighty kilometers. Subsequent work seems to put the height of the column at ten to sixteen kilometers.

Many believe that the Toba Volcano caused a volcanic winter over the globe. Scientists estimate that average worldwide temperature dropped between three to five degrees Celsius. The latitude for Toba Lake is approximately 3°, or nearly at the equator. Latitudes farther away from the volcano may have experienced drops of up to fifteen degrees Celsius. Ash deposits in the South China Sea indicate that the Young Toba Tuff probably erupted in the northern summer. Dust could only have reached this location because of the summer monsoons.

Then a few other less dramatic, but still quite devastating, events happened. Scientists purport that a meteor formed the Barringer Ranch Crater near Flagstaff, Arizona, fifty thousand years ago.[2] Scientists estimate the energy of the event at ten megatons of TNT. Even though the Barringer Ranch Asteroid was relatively small, scientists estimate that it still had massive effects on the environment.[3] The air blast was lethal to human-sized animals within a three-kilometer radius of the impact site. Five to six kilometers from the impact site, damage to living things occurred, making survival unlikely. Winds of 1,500 kilometers per hour at the impact site dropped to one hundred kilometers per hour, twenty-five kilometers from the impact site. Fifty kilometers from the impact site, damage to trees and human structures is moderate. Think of the effect a really large impact event would have.

---

[1] https://en. wikipedia. org/wiki/Lake_Toba
[2] https://en. wikipedia. org/wiki/Meteor_Crater
[3] D. P. Bobrowsky. *Comet and Asteroid Impacts and Human Society.* Ottawa/Uppsala, Ontario/Sweeden: Springer, 2007.

The Taupo Volcano erupted in New Zealand.[1] The volcano began eruptions eight hundred thousand years ago. The Oruanui eruption of the Taupo Volcano occurred approximately 26,500 years ago.[2] The Taupo eruption generated 430 cubic kilometers of solids that fell to Earth from its plume and 420 cubic kilometers of solids emitted as molten lava. The Oruanui eruption deposited an eighteen-centimeter-thick layer of ash in the Chatham Islands, some one thousand kilometers away.

Twenty-two thousand years ago, the Aira super volcanoes erupted in Japan. The Aira Volcanoes were equal to fifty Mount St. Helen volcanos, but much smaller than the Toba Volcano.[3] The date of the Aira Caldera fits the timeline, prior to the start of the last interglacial warming period.

Recent Events on Earth

A phenomenon, known as the "Older[4] and Younger Dryas[5]," caused the sub-anomaly during the last glacial warming period. The most popular theory attributes the Dryas to declining strength of the North Atlantic Ocean currents carrying warm water from the equator towards the North Pole. This allowed an influx of influx of fresh cold water from melting continental ice sheets, resulting in colder weather.

Another theory for the Dryas phenomena is the "impact hypothesis."[6] Scientists[7] believe that a storm of comet and asteroid strikes occurred 12,900 years ago. A report[8] authored by twenty-six scien-

---

[1] https://en. wikipedia. org/wiki/Taupo_Volcano
[2] https://en. wikipedia. org/wiki/Oruanui_eruption
[3] http://www. ranker. com/list/the-world_s-6-known-supervolcanoes/analise. dubner
[4] https://en. wikipedia. org/wiki/Older_Dryas
[5] https://en. wikipedia. org/wiki/Younger_Dryas
[6] https://en. wikipedia. org/wiki/Younger_Dryas_impact_hypothesis
[7] http://www. dailymail. co. uk/sciencetech/article-2158054/Scientists-discover-evidence-meteorite-storm-hit-Earth-13-000-years-ago-killed-prehistoric-civilisation. html
[8] http://www. pnas. org/content/104/41/16016. full

tists and noted authors[1] discusses how the discovered data reveals a period of devastation on Earth, just before the start of the six days of God's work. The storm caused fires that devastated the Americas and Syria. This caused smoke and soot to cover the globe. The findings of the report caused controversy in the science community. Some scientists had doubts about the theories presented, and some vehemently supported them.

As we will see in chapter 3, the Bible itself contains scientific details, although somewhat subtle, that support the impact theory discussed above. The Bible story points toward comet strikes as a major part of the storm.

The Rio Cuarto Comet that struck in Argentina is one event listed in scientific comet lists (Bobrowsky 2007). Scientists carried out little work on the Rio Cuarto Comet strike and so we do not know the exact strike date. Estimated dates range from six thousand to twelve thousand years ago. Nevertheless, the Rio Cuarto Comet, whether part of the comet and asteroid storm or not, is a good example of the devastation a comet can cause.

The impact of a collection of comets caused the Rio Cuarto craters in Argentina, shown in figure 2-7.[2] The collections of comets likely formed from a single comet that exploded before it struck the Earth. They formed ten craters on the Earth's surface, four of them of substantial size. One crater named the "Drop" was about two hundred meters wide and six hundred meters long. Two more large craters, the Eastern Twin and Western Twin, both about seven hundred meters wide and 3.5 kilometers long, were located five kilometers to the northeast. Another major crater, the Northern Basin, is about half again as big as one of the Twins. Scientists sited the Northern Basin eleven kilometers further to the northeast. The long axes of the craters all point to the northeast.

The Chelyabinsk meteor in Russia travelled at a speed of sixty thousand to sixty-nine thousand kilometers per hour. It was twenty meters in diameter and had a mass of twelve thousand to thirteen

---

[1] R. B. Firestone. Evidence for an extraterrestrial impact 12,900 years, 2007.
[2] https://en. wikipedia. org/wiki/Rio_Cuarto_craters - By NASA - Public Domain

thousand metric tonnes. Scientists estimate the Chelyabinsk meteor that struck in Russia on February 15, 2013, had an energy of five hundred kilotons of TNT.[1] This was twenty to thirty times the energy of the atomic bomb dropped on Hiroshima, Japan. Scientists estimate that the Rio Cuarto comet was ten times[2] the size of the Barranger Ranch Asteroid, or one hundred megatons of TNT. This is equivalent to six thousand Hiroshima bombs (Barrientos 2012). Some studies state the Rio Cuarto Comet as high as fifty thousand to five hundred thousand Hiroshima bombs.[3]

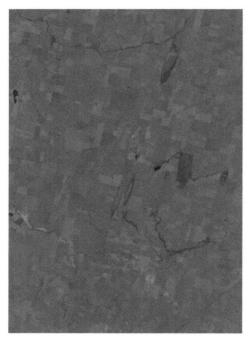

Figure 2-7- Rio Cuarto comet strike craters

The Rio Cuarto comet approached the Earth at an angle of fifteen degrees to the Earth's surface and flew near the Earth's surface over a long portion of its northeasterly to southwesterly flight path. A huge pressure wave possibly created a tsunami off the coast of South America. The pressure wave pushed huge volumes of dust into the air as it passed over the Sahara Desert in Northern Africa. When the comet struck the Earth, huge amounts of dust and debris entered the Earth's atmosphere as it created huge craters on the Earth's surface. Water and other debris rained down on the Earth from the comet's tail after impact.

---

[1] https://en. wikipedia. org/wiki/Chelyabinsk_meteor
[2] https://en. wikipedia. org/wiki/Rio_Cuarto_craters
[3] Barrientos, G. -M. (2012). The Archaeology of Cosmic Impact: Lessons from Two Mid-Holocene Argentine Case Studies. Springer Science+Business Media.

I generally reviewed available data from various sources. Figure 2-5 shows a very rough estimate of what the Dryas temperature anomalies may have looked like (in brown).[1]

Science Proves the Bible

Establishment of the time in history of events in Genesis clears up many of the false beliefs, advocated by both supporters and detractors of the Bible. Fixing Genesis in time removes the arguments of whether the detailed narrative of Genesis includes the formation of the universe or not. As we see, it does not. Although chapter 2 of Genesis briefly mentions the formation of the universe, nearly all the book covers a time billions of years later.

About fifty thousand years before the start of God's work, a devastating mass extinction, the Quaternary Mass Extinction, started. Subsequent catastrophic events put more pressure on the struggling living things that survived the Quaternary Mass Extinction. The mass killing of many plants, animals, and early Homo sapiens by the Quaternary extinction event itself reduced the numbers on the Earth. Very many species became extinct. Then survivors had to endure super volcanoes and celestial body strikes. These events blanketed the globe with dust and depleted the food supply. The explosive nature of these events likely wiped out much life directly, not to mention the lingering aftereffects on the food and water supply. Any surviving living things probably looked haggard, to say the least. Dust from volcanoes and celestial body strikes spread around the Earth, up to several meters deep in places. This very likely resulted in dust storms as the wind picked up and moved the dust around the globe.

The "formless and empty" conditions in the Bible certainly reflect what science tells us about past catastrophic events on Earth.

---

[1] Base Ice Core Data taken from World Data Center for Paleoclimatology, Boulder and NOAA Paleoclimatology Program, AICC2012 800KYr Antarctic Ice Core Chronology, Original_Source_URL: ftp://ftp. ncdc. noaa. gov/pub/data/paleo/icecore/antarctica/aicc2012vostok. txt

3

Days 1–4: Earth Recovers

Contemporary religions teach that days 1–4 of God's work involved the creation of the Earth, sun, moon, and stars, in that order. This completely flies in the face of science. Detractors often use this as a basis to discredit the Bible. As we will see, the real story of the Bible matches scientific facts.

Chapter 2 showed how the Earth became devastated by a series of catastrophic events. In this chapter, we will see that days 1–4 of God's work involved no creation. We will see how the Bible tells us of a time that the Earth recovers from the results of the previous devastating events.

Comet Strikes—A Key Factor

Comets produce great changes on the Earth and in its atmosphere. Comets continually pass by the Earth, as shown in Figures 3-1[1] and 3-2.[2] Comets originate outside the solar system.

Scientists[3] have learned from spacecraft missions that the composition of different comets varies. The comet Halley's composition is 80

---

[1] https://en. wikipedia. org/wiki/Comet- Hubble Image by European Space Agencyis licensed under Creative Commons Attribution 4. 0 Unported license
[2] https://en. wikipedia. org/wiki/Comet- Dust Trail by NASA is licensed under NASA rules in the Public Domain
[3] *Deeppace.* New York: Cambridge University Press.

percent ice, a smaller amount of gases such as carbon monoxide, and a small amount of dust. Impact experiments with the Temple 1 comet released material from the comet's nucleus that reportedly consisted of only five million kilograms of water, a surprisingly high ten to twenty-five million kilograms of dust. The composition of comets varies widely.

The measurement of a comet's size is a tricky business. Comets come in various sizes and shapes. The Temple 1 comet is approximately 7.6 kilometers × 4.9 kilometers and has a mass of 75 trillion kilograms (Eicher 2013). Other sources[1] estimate comet sizes that vary from one hundred meters to over thirty kilometers.

Comets approach the solar system as frozen blocks of ice and other matter. As the comet approaches the sun, a coma forms around the comet's nucleus as water begins to sublimate directly into a gas. Sublimated water along with dust in the comet forms a coma that surrounds the nucleus. A large halo of hydrogen makes up the outer layers of the coma. As the comet gets closer to the sun and continues to warm, the volume of its coma increases as more gases and dust burst forth. The sun's radiation and solar wind drives matter from the coma into a huge tail. Water and

Figure 3-1 Hubble image of Comet ISON shortly before perihelion - ESA/Hubble

Figure 3-2 Diagram of a comet showing the dust trail, the dust tail (or antitail) and the ion gas tail, which is formed by the solar wind flow - NASA

---

[1] https://en.wikipedia.org/wiki/Coma_(cometary)

gases form a plasma tail that points directly away from the sun. Often dust particles form a separate tail.[1]

The comet's coma grows as the comet approaches the sun and can be thousands to millions of kilometres in diameter. The ionized tail of the comet can be up to 150 million kilometers long. The dust released from the coma also forms a tail. Gases leaving the comet's nucleus often form a jet, which may cause the comet to spin and even explode.[2]

Scientists tell of two parts of the coma of comet Halley (Swamy 2010). A visible part of the coma is ten thousand to one hundred thousand kilometers in size, while an invisible, ultraviolet part is one million to ten million kilometers in size. The dust tail of the comet is approximately ten million kilometers long.

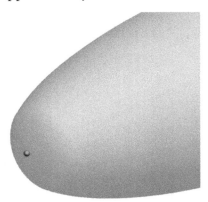

Figure 3-3 Comet size relative to Earth

Figure 3-3 depicts a comet approaching the Earth. The dimensions of the comet's coma shown are for a small to midsized comet, with a coma less than 350,000 kilometers in diameter. Figure 3-3 is to scale. The Earth is the very small sphere, shown in blue at the front of the comet's coma. The materials in the comet's coma will not only envelope the Earth's atmosphere but also a very large area of space around the Earth. Clearly, a comet strike will change the environment in space around the Earth for a significant distance and time period. After the comet strikes, the Earth's gravity will attract debris in the comet's tail. The sun's gravity will also attract the tail, causing the tail to orbit the sun with the Earth.

As discussed previously, various theories exist regarding how the comets and other events such as asteroids, volcanoes, and the Dryas affected conditions on Earth. The Bible's description of the Earth at

---

[1] *Deeppace.* New York: Cambridge University Press.
[2] http://en.wikipedia.org/wiki/Comet

the beginning of day 1 of God's work certainly suggests that comets, such as the Rio Cuarto Comet, or asteroids played an important role.

Comets travel at very high velocity. The moving debris from the comet's tail will strike and rush past the Earth, causing great turbulence. Figure 3-4 shows the effect of the tail of a comet rushing past the Earth. The high velocity of matter in the comet's tail causes huge storms on the face of the Earth. This displaces the atmosphere on the side of the Earth facing the approaching comet to the other side of the Earth. A sudden loss of oxygen results on the side of the Earth facing the comet.

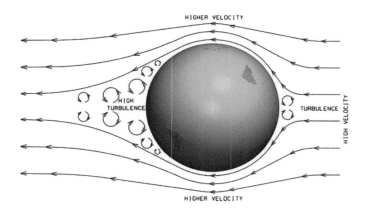

Figure 3-4 - Turbulence from a comet and displacement of the Earth's atmosphere

Day 1: Light Appears

I lived on a prairie farm in Saskatchewan. I can still remember old-timers talking about the dirty thirties. Settlers on the plains fought with daily black clouds of dust that rolled over the land. The dust penetrated every crack and settled on tables and floors, sometimes in wavelike patterns. The dust storms literally turned day into night. The conditions during the dirty thirties give us some inkling of conditions on the Earth thirteen thousand years ago, when the six days of God's work started. The dust during the dirty thirties was due to wind blowing eroded soil from poor farming practices. Can you even imagine the conditions caused by the events before the start of God's six days of work? Before the start of the six days of God's work, volcanic dust covered the whole globe. The dust layer was up to several meters deep in some parts of the world. Violent windstorms resulted from the temperature variations, common in glacial periods. The wind blew the volcanic dust around the globe. Then several celestial body strikes covered space around the Earth with a mixture of water and debris.

Figure 3-5 illustrates what the Earth may have looked like because of the cataclysmic previous events.

Figure 3-5 Darkness at the start of day 1

The dust and water in the atmosphere completely blocked the sun. Campfires and open outdoor hearths made the only light, which only illuminated nearby objects. Animals and early man had to grope around in this dark environment to search for anything that might serve as food. Dirty water not fit for human consumption covered the Earth. Nevertheless, all living things had to drink liquid that looked more like effluent from a sewer than water. Death, misery, and despair affected the general mood of all living things. The Earth's occupants had to exist on higher land, above the slimy water that covered most of the Earth. Here, what little vegetation that survived provided them a bare existence, which only allowed the most fit to survive.

THE REAL BIBLE AND SCIENCE

> And God said, "Let there be light," and there was light. God saw that the light was good, and he separated the light from the darkness. God called the light "day," and the darkness he called "night." And there was evening, and there was morning—the first day.[1]

These words describe God's work on the first day. The original Hebrew Bible wording for the term "let there be" comes from a word that implies "come to pass" or "become." The word "become" is especially telling when we look at the original Hebrew word for the English word "light." The Hebrew word implies joy. The establishment of light changed the dreary, sorrowful state of the world to one of more cheerfulness. Things were still not great for living things. The dim light made things less obscured than before. Plants tried to start growing again. The remnants of plants not completely dead showed signs of life. Things started to look up, and there was reason for happiness because the worst seemed over, and improvements were on the horizon. Comet and asteroid strikes blocked the sun's light and created darkness over the Earth. To occupants of the Earth, the sun had disappeared. However, God did not have to create the sun to solve the problem.

The LORD already brought about the sun after the big bang, billions of years earlier. God's words tell of the existing sun's rays starting to reach the Earth. Laws of gravity, which were part of the LORD's original formation of the universe, were already in place. When these laws took effect, gravity pulled larger dust particles and debris toward the Earth. This partially cleared the atmosphere, letting a very small amount of diffuse sunlight light through to Earth.

The passage ends with "God called the light "day" and the darkness he called "night." And there was evening, and there was morning—the first day." This passage shows us that after the first day, God established day and night. The clearing of dust from the atmosphere let light from the sun to the Earth's surface. Obviously, the Earth

---

[1] Genesis 1:3

was rotating, thereby creating the night and day we know today. The Hebrew word for "called," means "to decree" or "to declare," not "to create." As the original

Figure 3-6 Light at the end of day 1. Very foggy, flooded land near ocean

The Bible indicates, no act of creation occurred.

Figure 3-6 shows a picture of what the end of day one may have looked like on Earth. There is still much of water vapour in the air. Only nearby landmasses are visible. Dirty water still covers the Earth.

Earth's Atmosphere and Magnetic Field

Detractors often have huge negative reactions to the Bible story of God's second day of work. They feel that the separation of water from water in the Earth's atmosphere makes absolutely no sense. But the Bible is correct.

Some authors call the separation of waters a myth and scientifically unacceptable.[1] Bucaille probably overlooks two important facts. The first is the Earth's magnetic field. The second is the fact that a comet's tail contains ionized matter, or a plasma, which surrounds the Earth during a comet strike.

We need to understand some basics about the Earth's atmosphere and magnetic field to make sense of the Bible's story about day

---

[1] D. M. Bucaille (n. d.). The Bible The Qur'an and Science.

2 for God's work. Thomas Hill and Richard Wolf[1] define the Earth's magnetic field as the outside the Earth's environment in space. The Earth's magnetic field extends over sixty-four thousand kilometres up from the Earth's surface, on the side of the Earth facing the sun and 1.3 million miles on the side of the Earth facing away from the sun.[2] The lower limits of the magnetosphere[3] starts between ninety and 120 kilometers above the Earth's surface and is a space where the Earth's magnetic field governs the actions of ions and electrons.[4]

The Earth's magnetic field protects the Earth from the solar winds, which are high velocity, charged particles from the sun. The magnetosheath extends past the Earth's magnetic field. The magnetosheath is where the Earth's magnetic field slows down and redirects the solar wind. This diversion of the solar winds prevents the solar winds from blowing away the Earth's atmosphere into space.

The sun's solar winds sometimes produce a plasma that can seriously disrupt the Earth's magnetic field because a plasma is a very good electrical conductor.[5] The density of the solar winds is 10^{-19} (one with nineteen zeroes in front it) that of the atmosphere

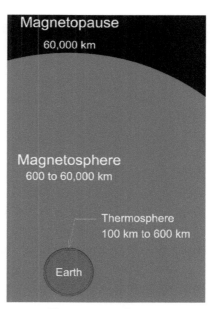

Figure 3-7 – Lower levels of inner space

---

[1] T. W. Hill. *Solar Wind Interactions*. Washington, DC, USA: National Academy of Science, 1977.
[2] J. Baggott. *Origins - The Scientific Story of Creation*. Oxford, Unite Kingdom: Oxford University Press, 2015.
[3] J. L. Burch. *The Magnetosphere*. Washington: National Academy of Sciences, 1977.
[4] J. L. Burch. *The Magnetosphere*. Washington: National Academy of Sciences, 1977.
[5] Merril, R. T. (1996). *The Magnetic Field of the Earth*. London, UK: Academic Press.

on Earth[1] (Hill 1977). A comet and its coma and tail have densities higher than the Earth's atmosphere. Thus, a comet will disrupt the Earth's magnetic field immensely as it penetrates the Earth's atmosphere and crashes to Earth. A large comet will likely totally overwhelm the Earth's magnetic field.

Figure 3-7 shows the magnetosphere down to the thermosphere, the topmost layer of the Earth's atmosphere. Near the bottom of the thermosphere and higher, molecules are so far apart they rarely collide with one another, and the Earth's magnetic field only affects and controls them. In this region, gases stop acting like normal gases, like they do in the lower atmosphere. This occurs for electrons over ninety kilometers and ions over 120 kilometers above the Earth's surface[2]. Below this level, gases are denser and behave like normal gases behave on the Earth's surface.

Figure 3-7 provides an indication of the immensity of the magnetosphere and space. The space station orbits 330 to 435 kilometers above the Earth's surface in the thermosphere.

Figure 3-8 shows layers of the atmosphere closest to Earth. The mesosphere exists below the thermosphere. Temperatures as low as minus 143°C are possible in the mesosphere.[3] The mesosphere sometimes contain noctilucent[4] cloud, the highest-level clouds above Earth. Spacecraft fly above the mesosphere and aircraft below the mesosphere, and so very little data exists for this level.

The ozone layer exists approximately twenty kilometers above the Earth's surface.[5] It varies in depth and can exist as low as twelve kilometers to twenty-five kilometers above the Earth's surface. The sun's radiation reacts with oxygen to form single atomic oxygen (O) and ozone (O3). The ozone layer is vital to life on Earth. It stops dangerous ultraviolet light from the sun from reaching the Earth.

---

[1] Hill, T. W. (1977). *Solar Wind Interactions*. Washington, DC, USA: National Academy of Science.
[2] Burch, J. L. (1977). *The Magnetosphere*. Washington: National Academy of Sciences.
[3] https://en. wikipedia. org/wiki/Mesosphere
[4] https://en. wikipedia. org/wiki/Noctilucent_cloud
[5] B. J. Pitts. *Upper and Lower Atmosphere*. Irvine, California, USA: Acedemic Press, 2000.

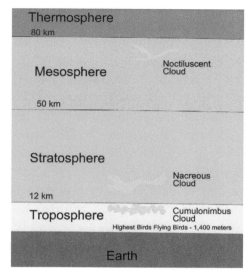

Figure 3-8 Levels of atmosphere Near Earth

The troposphere[1] exists next to the Earth. It is the part of the atmosphere where we live. Nearly all the Earth's weather phenomena take place in the troposphere. It contains 75 percent of the total atmosphere's mass and 99 percent of all atmospheric water.

Cumulonimbus[2] clouds form in the troposphere and cause thunder storms and instability in the atmosphere. Many other types of cloud exist at elevations below and above the altitude of the clouds shown in figure 3-8.

Day 2: The Separation of Waters

> And God said, "Let there be a vault between the waters to separate water from water. So God made the vault and separated the water under the vault from the water above it. And it was so.[3]

Modern translations do not quote this verse correctly. The original Hebrew Bible reads similar to "God made a firmament, he separated between the waters which were under the firmament, from the waters which were above the firmament."

Before day two, incoming comets saturated space around the Earth, starting at the Earth's surface and including the magneto-

---

[1] https://en.wikipedia.org/wiki/Troposphere
[2] https://en.wikipedia.org/wiki/Cumulonimbus_cloud
[3] Genesis 1:6-7

sphere and beyond, with a thick mixture of ionized water and dust. The enveloping cloud overwhelmed the Earth's magnetic field. There was little difference between atmosphere at the Earth's surface and far out into space. By day 2, the thick layer started to thin out because the Earth's gravity pulled much of the surrounding water and dust to Earth. The density of the magnetosphere again became very low. The Earth's magnetic field started to regain its strength. This caused gases in the magnetosphere to start to behave in the same manner as before; the gases did not behave as gases on Earth and became mainly affected by the Earth's magnetic field only. The magnetosphere now again repelled water plasma and other material from space, just as it does today. This protected the Earth's atmosphere, or the troposphere, from incoming water and other matter from space. As the Bible states, water was separated from water.

Thus, after the second day, or two thousand years after the start of God's work, a firmament separated the Earth's near atmosphere from outer space. The Bible's description of day 2 makes complete sense when examined with an accurate definition of what science tells us about the Earth's magnetic field and space around the Earth. The critics' detractions evaporate with this new knowledge.

Figure 3-9 - The atmosphere is less foggy at the end of day 2, but water covers land near ocean.

Figure 3-9 depicts the end of day 2. Although still foggy, distant mountains and the far horizon are now visible. The Earth is still waterlogged. A distant mountain past the horizon becomes visible.

Day 3: The Sky Clears and the Land Dries

> And God said, "Let the water under the sky be gathered to one place, and let dry ground appear." And it was so. God called the dry ground "land," and the gathered waters he called "seas." And God saw that it was good.[1]

Water in the troposphere now gathered into one place. Some water remained in the atmosphere as clouds and the rest fell to Earth. The original Bible text tells of a slow, long-lasting process, where the waters lingered a long time before disappearing into the sea. The appearance of "dry ground" implies that ground was already visible on the Earth, but the ground was wet. The Hebrew word for "dry ground" implies that the ground was a wretched disgrace in its wet state. This fits the description of "void" in the second verse of Genesis, where the Earth appears as a watery wasteland.

Figure 3-10 shows day three. The illustration depicts some time before the end of day three. Dry land has formed. God brought about vegetation on day three. Figure 3-10 shows the scene before vegetation started growing.

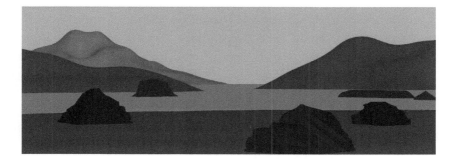

Figure 3-10 Day 3: Things continue to clear up. Land near ocean is separate from ocean and dry.

---

[1] Genesis 1:9

Day 3: Plants Recover

> Then God said, "Let the land produce vegetation: seed-bearing plants and trees on the land that bear fruit with seed in it, according to their various kinds.[1]

As with the stars, sun, moon, and Earth, Genesis does not say that God created plants. The Hebrew word for the English word "produce" means to emerge with a connotation of germinate or grow. Here again, the Bible implies not an act of creation but one of emergence.

The original Hebrew text mentions three special seed-bearing types of plants only—grass, herbs, and trees. The original Hebrew text tells us that the plants brought forth on day 3 were a special kind of plant that humans could use for food. The grass family includes cereal crops such as modern-day wheat, oats, and barley, which is food for man. As well, grass is food for animals, which in turn provide food for man. We use many herbs today in cooking and for medicinal purposes. The original Bible text describes fruit trees that have seed in the fruit. This includes all kinds of plants including modern-day apples, pears, cherries, and berries as well as trees that produce nuts. The original Bible contains a phrase at the end of the above verse not shown in contemporary translations. The phrase says, "Whose seed, in itself, is against (on) the Earth." The plants in question produce their offspring with seeds that are on and against the earth.

The Bible then goes on to tell us that the seed on the ground produced its bounty: "And it was so. The land produced vegetation: plants bearing seed according to their kinds and trees bearing fruit with seed in it according to their kinds."[2]

But what about seed-bearing plants that do not produce fruit and non-seed-bearing plants? Were other types of plants already in

---
[1] Genesis 1:11
[2] Genesis 1:12

existence? Scientists say yes. In chapter 7, I show how plants were in existence billions of years earlier.

Around 12,500 years ago, temperatures in Europe rose by at least 6°C in the summer.[1] About twelve thousand to nine thousand years ago, grassland common to colder temperatures, flowering grass plants and herblike grasses dominated the vegetation. A gradual spread of forest-like vegetation started ten thousand years ago, starting in river bottomlands and foothills. Initially, pine and birch trees started growing, followed by broad leaf species such as oak, elm, and a type of hazelnut tree. Initially plants requiring minimal amounts of water grew. Plants and their seed definitely existed, prior to day 3. Science tells us that growing conditions improved. This allowed existing plant seeds to germinate and existing plants start growing more vigorously.

The Hebrew word for "land" means productive land like that found in a farmer's field. Obviously, the plants included are domestic plants, not wild plants.

In chapter 1, I showed that chapter 2 of Genesis flashes back to chapter one of Genesis. The reestablishment of plants on the Earth is a subject that ties chapters 1 and 2 of Genesis together in a most definitive way. Chapter 1 states, "Then God said, 'Let the land produce vegetation: seed-bearing plants and trees on the land that bear fruit with seed in it, according to their various kinds.'"[2] Chapter 2 states, "Now no shrub had yet appeared on the earth and no plant had yet sprung up."[3] Chapter 2 obviously describes a time before the appearance of plants in chapter 1.

The Hebrew words in the original Bible indicate shrubs and plants grew around or in fields or agricultural land. The Hebrew word used for plant could include herbs, but could also refer to grass or cereal crops. Undoubtedly, chapter 2 of Genesis takes us back to chapter 1 after seeds fell to the ground, but before they sprouted or germinated.

---

[1] Springer. *The Black Sea Flood Question:*. Dordrecht, The Netherlands, The Netherlands: Springer, 2007.
[2] Genesis 1:11
[3] Genesis 2:5

Day 4: The Mist Clears and the Sun Comes Out

> And God said, "Let there be lights in the vault of the sky to separate the day from the night, and let them serve as signs to mark sacred times, and days and years, and let them be lights in the vault of the sky to give light on the Earth."[1]

The original Bible text uses only two words for the English words "vault of the sky." The first word for vault is "firmament," which refers to all and even the farthest reaches of outer space. The second word, "heavens," refers to the highest heavens or what one sees when looking up at the sky on a clear night—in other words, the universe.

> God made two great lights—the greater light to govern the day and the lesser light to govern the night. He also made the stars. God set them in the vault of the sky to give light on the Earth, to govern the day and the night, and to separate light from darkness. And God saw that it was good. And there was evening, and there was morning—the fourth day.[2]

The Hebrew word for the English word "made" has a different meaning than the word "made" in the contemporary English language. The Hebrew word used for the word "made" means to assign, select, or confer. The original Hebrew word used for the English word "set" means to confer, delegate or accredit. Thus, God selected the already existing sun, moon, and stars and assigned them the duties described in the Bible. There is no mention of creation. Finally, after four thousand years, the sun, moon, and stars become visible. Figure 3-14 depicts the end of day 4. The sun is now out, and trees and grass have grown.

---

[1] Genesis 1:14–15
[2] Genesis 1:16–19

Figure 3-11 Day 4: The sun reappears

The Bible and Science Agree

The Bible's story of days 1–4 removes the supposed conflicts detractors say exist with science. We see that the sun, moon, and stars were in existence before days 1–4. The creationists' view is wrong. The Earth was not in existence before the sun, moon, and stars, as they propose. We also see that no creation happened on days 1–4. The wording of the Bible tells us that God only decreed that certain types of plants progress forward—plants that serve as food for man and animals. The wording of the original Bible tells us that the sun, moon, and stars already existed. Their reappearance required only a clearing atmosphere.

4

The Start of Civilization

My early religious education taught me that Adam was the first man on Earth. We will see that man existed on Earth many years before Adam. Science tells us that the tragic events preceding God's work greatly affected these humans and the way they lived. We will see that God interacted with these early humans. They apparently would not work with God, so the LORD God formed Adam to fulfill his requirements.

After his initial creation, the LORD put Adam in the Garden of Eden. We will see how a correct translation of the original Hebrew Bible allows us to locate the Garden of Eden. With the help of the Bible and scientific facts, we will see that the Garden of Eden story could only happen on day 3 or 4, not on day six as many religions teach.

After God drove Adam and Eve out of the Garden of Eden, Eve gave birth to Cain and Abel. Cain killed Abel, so God drove Cain away from his presence. We will see how the Bible reaffirms the existence of other humans on Earth, in this part of the Bible story.

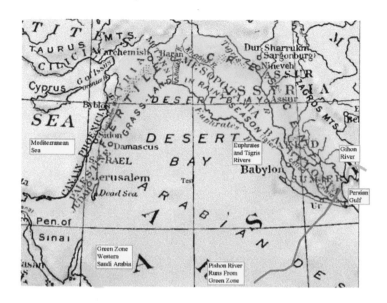

Figure 4-1- The Fertile Crescent

Earliest Man Preexisted God's Six Days of Work

Chapter 7 will provide more details of how man evolved on Earth. Science tells us that man existed on Earth long before the time of Genesis. Sources indicate that anatomically modern humans replaced the Neanderthals, approximately forty thousand years ago.[1]

The story of Genesis takes place in and around the ancient land called Mesopotamia. Mesopotamia means "the land between the rivers."[2] Figure 4-1 shows the Fertile Crescent, a part of Mesopotamia known for its lush fertile land during all periods of history. The Fertile Crescent is a crescent-shaped stretch of land starting at the Persian Gulf.[3] It continues northward from the Persian Gulf and runs through present-day northern Syria, southern Turkey, and through

---

[1] Springer. *The Black Sea Flood Question:*. Dordrecht, The Netherlands, The Netherlands: Springer, 2007.

[2] K. R. Nemet-Nejat. *Daily Life in Ancient Mesopotamia*. Westport, CT, USA: Greenwod Press, 1998.

[3] https://en. wikipedia. org/wiki/Fertile_Crescent

Iran and Iraq. The Fertile Crescent continues down along the east side of the Mediterranean Sea and into Egypt.

The Euphrates, Tigris, Gihon Rivers, and previously the Pishon River run through the area and are the source of life for inhabitants. The Pishon River has dried up in the distant past and is no longer on modern maps. The location shown in figure 4-1[1] is approximate.

El Kowm,[2] Syria, shown in figure 4-2,[3] is the home of archaeological remains of the first Homo erectus. The discovered Homo erectus lived approximately 450,000 years ago. The location of El Kowm is on an oasis in the desert, outside the Fertile Crescent. Science tells us that two hundred thousand years ago, Homo erectus evolved into Homo sapiens, the species that modern man belongs to.

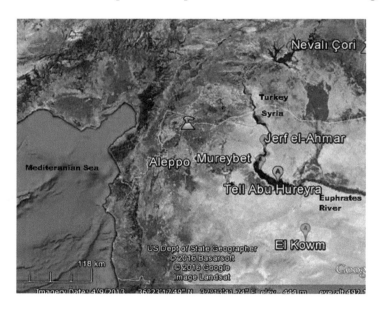

Figure 4-2- Early settlements along the Northern Euphrates River

---

[1] https://upload. wikimedia. org/wikipedia/commons/2/22/Fertile_Crescent_concept_1916. png- Figure taken from this source and added to – Public Domain

[2] https://en. wikipedia. org/wiki/El_Kowm_(archaeological_site)

[3] Google Earth Maps – Image altered for this book – Used per Google terms https://www. google. ca/permissions/geoguidelines. html

Iraq and Syria formed a former crossroads between Africa and Eurasia.[1] The El Kowm Oasis in central Syria contains artefacts that show human history from almost 1.8 million years ago to the present. Early man belonged to the hunter-gatherer culture. He wandered the Earth and lived on whatever food the Earth provided. Early man did not form communities, and his culture probably had very little, if any, formal social structure.

Archaeologists refer the period from seven hundred thousand years ago to approximately eleven thousand years ago as the Paleolithic or Old Stone Age.[2] Inhabitants used only stone tools and no villages existed. Eleven thousand years ago, at the start of the Neolithic (New Stone) Age, the world's first permanent settlements of up to ten hectares with populations of up to one thousand or more appeared.

The events preceding God's six days of work affected the formation of permanent settlements, more than anything else. The nomadic hunter-gatherer way of life became unliveable. The catastrophic events before God's six days of work and the rain of watery debris from the sky made survival much more gruelling. The conditions made travel tenuous. People stayed in one place. As we will see, these settlements started in the upper reaches of the Fertile Crescent,[3] along the valley of the Euphrates River. This region was in higher elevations, above the lower plains where water and mud from the sky tended to accumulate.

Earliest Settlements Show Advancing Bad Conditions

The time early settlements started suggests an effort by residents to deal with the events that made the Earth formless and empty. Scientists tell us that some of the first settlements began at Tell Abu

---

[1] R. E. L. Jagher. El Kowm Oasis, human settlement in the Syrian Desert. Science Direct, 2015.
[2] K. R. Nemet-Nejat. *Daily Life in Ancient Mesopotamia*. Westport, CT, USA: Greenwod Press, 1998.
[3] https://en. wikipedia. org/wiki/Euphrates

Hureyra.[1] [2] This is the most southern of a series of settlements, as shown in figure 4-2. The settlement started about thirteen thousand years ago, spanning the cool, dry Younger Dryas, before God's six days of work. Scientists refer to this era as the "late Pre-Pottery Neolithic and the Pottery Neolithic." Circular mud brick huts, crop cultivation, and the domestication of cattle, sheep, goats, and pigs started. Hunting declined, and pottery emerged.

Inhabitants abandoned Tell Abu Hureyra twelve thousand years ago, just as the first day of God's work started. A subsequent rehabitation started 10,400 years ago and lasted until 7,800 years ago. Scientists think the abandonment was due to a change to colder weather, due to the Younger Dryas. Colder weather probably exacerbated problems by causing water and mud to freeze on roofs, possibly causing buildings to collapse. The location of Abu Hureyra was on a low table protruding into the Euphrates River. The elevation of Abu Hureyra was as low as one possibly could get.[3]

Scientists tell us that 11,700 years ago residents inhabited Jerf el-Ahmar.[4] Jerf el-Ahmar was like Tell Abu Hureyra, with circular huts evolving to rectangular buildings.[5] Around 10,000 to 9,400 years ago construction changed. Larger rooms indicate the possibility of some sort of communal activity. As the author states, specialized buildings, such as the "house of the bulls" seemed to imply a ritualistic standing with aurochs symbolizing special status. Wild barley, einkorn, and rye chaff indicate primitive forms of crop use. Jerf el-Ahmar lasted only until 10,500 years ago.

---

[1] https://en. wikipedia. org/wiki/Tell_Abu_Hureyra
[2] J. Ridout-Sharpe. Changing lifestyles in the northern Levant: Late Epipalaeolithic and early Neolithic shells from Tell Abu Hureyra. Quaternary International, 2015.
[3] J. Ridout-Sharpe. Changing lifestyles in the northern Levant: Late Epipalaeolithic and early Neolithic shells from Tell Abu Hureyra. Quaternary International, 2015.
[4] http://archive. archaeology. org/0011/abstracts/farmers. html
[5] A. H. Simmons. *The Neolithic Revolution in the near*. University of Arizona Press, 2011.

People started the settlement of Mureybet[1] 10,500 years ago.[2] Archaeological excavations show Mureybet on a six-meter-high by seventy-five-meter round mound. The mound sits on a four-meter-high, 125-meter-wide by 250-meter-long raised area, adjacent to the Euphrates River. The site is three hundred meters above sea level and as high as twenty meters above the Euphrates River. The area consisted of flint embedded in chalk cliffs. Inhabitants obviously did not select the area as prime farmland. Residents clearly sought higher ground in the face of water and debris raining down from the sky. Most buildings hugged the top of the mound. Clearly protection against rising waters took priority over selecting land good for production of crops.

At Mureybet, initially huts only 1.5 meters in diameter sufficed. The Residents made even stronger huts, by reinforcing walls with embedded stones.[3] By 8,300 years ago interior diameters grew to six meters, with buildings excavated into the soil. A light wood structure supported roofs of brush or foliage. Later houses dropped to four meters in diameter, with surface construction dominating. By eight thousand years ago, the number of stone sickle blades starts to increase. The number and quality of scrapers and arrowheads also increase, suggesting an increase in hunting. By 7,300 years ago, rectangular houses appeared. Wall construction improved to rough limestone bricks or loaves with thick red clay mortar. Houses were approximately 3.5 meters square, with four 1.5-meter square rooms. Signs of ritual behaviour appeared. During the period 8,900 years ago, the population of Mureybet declined.

---

[1] https://en. wikipedia. org/wiki/Mureybet
[2] C. K. Maisels. *The Emergence of Civilization*. New York, NY, USA: Routledge Inc., 1990.
[3] J. Ridout-Sharpe. Changing lifestyles in the northern Levant: Late Epipalaeolithic and early Neolithic shells from Tell Abu Hureyra. Quaternary International, 2015.

Archaeological data indicates that residents started the settlement of Nevalı Cori[1] 10,200 years ago.[2] Nevali Cori is in Turkey, northeast of the Mediterranean Sea. Settlements to the south were at elevations in the range of two hundred to three hundred meters above sea level. The elevation of Nevali Cori was 490 meters above sea level and nestled in the foothills of the Taurus Mountains. Nevali Cori was well above the elevation of previous settlements and most able to withstand the effects of run off towards the oceans. The record indicates that buildings in Nevali Cori date back to 9,700 years ago. Buildings appear the most advanced in Nevali Cori, advancing quickly from round construction to rectangular construction. Larger buildings apparently served as some form of cult worship centers. In Gobekli Tepe, adjacent to and immediately south of Nevali Cori, residents built large temples, using pillars six meters high and weighing twenty tons each.

The continual movement to higher and higher ground indicates that residents sought to secure sites above the elevation of rising water and mud. The number of sickle tools suggests cultivation appeared, mostly likely because of wild grain growing in the area. Perhaps cultivation was a way to expose wild plants covered by the debris raining down from above. Light started to appear on day one, making plants more likely to grow faster, if exposed. The improvement in building construction over time suggests residents continually battled with the debris raining down from above. They had to increase the strength of buildings, to withstand the weight of the debris raining down from above.

God Intervenes to Improve Man's Existence

In chapter 3, we saw how God sought to promote agricultural plants on day three. A more detailed review of chapter 2 of Genesis tells us how God planned to do this, as discussed below. The Bible goes on to describe two impediments to God's plan: "Now no shrub

---

[1] https://en. wikipedia. org/wiki/Neval%C4%B1_%C3%87ori
[2] M. Tobolcyk. *The World's Oldest Temples in Gobekl and Nevali Cori.* Warsaw: Warsaw University of Technology, 2016.

had yet appeared on the earth and no plant had yet sprung up, for the Lord God had not sent rain on the Earth and there was no one to work the ground."[1] As discussed previously, this verse tells of a time when seeds lay on the ground, but before they sprouted and took root in the soil.

The verse goes on to give two reasons why plants had not yet started to grow. The first reason is lack of rain. In chapter 2, we saw that the Younger Dryas was active during this time, and Mesopotamia had no rain during this period. Science tells us that conditions were arid because ocean levels were low, prior to the melting of the glaciers. Studies (Hughlett 2016) indicate that temperature and precipitation started to increase 11,500 years ago and began to reach more normal levels about ten thousand years ago, during day 3 of God's work. The Bible agrees with science.

The second reason that plants would not grow is that there was no one to work the ground. The next verse of chapter 2 is highly misunderstood by most people, including myself until I wrote this book: "But streams came up from the earth and watered the whole surface of the ground."[2] This verse seems to imply that water came up from the earth and watered the ground, to make up for the absence of rain. The original Hebrew text reads similar to "a haze came up from a surface, which swamped the ground below."

The Hebrew word for the English word "surface" means "face of" or prior to, in the sense of conflict or anger. The word implies hardness or strength. The surface of the land fought with a strong, hard adversary on top of it. The Hebrew word for "watered" means "to oblige," "to swallow," or "to drink," with a connotation of swamping or drowning. While this adversary emanated a fog or mist into the air above it, it also overcame the soil below it with too much water.

While pondering this apparent contradiction, I came across an article[3] (Fasihnikoutalab 2015), who discusses how the mineral olivine can stabilize soils instead of cement. In some cases, civil engineers

---

[1] Genesis 2:6
[2] Genesis 2:8
[3] M. H. Fasihnikoutalab. *New Insights into Potential Capacity of Olivine in Ground Improvement.* Department of Civil Engineering, University Putra Malaysia, 2015.

stabilize soil by mixing cement with the soil. The cement hardens in the soil, making the soil much stronger and harder. The article describes how Olivine[1] strengthens and hardens soil just like cement. Olivine forms silicon dioxide as part of the reaction when mixed with soil and water. This exothermic reaction[2] releases heat.

Comets and asteroids contain significant amounts of olivine[3][4] [5][6] (Eicher 2013) (de Vries 2012). This discovery puts the Bible text into perspective. The Rio Cuarto Comet or other comets like it deposited olivine on the Earth's surface. Just like cement, olivine hardened and formed a hard layer over the surface of the land. Comets also deposit water on the Earth's surface. The watery mixture of olivine and water swamped the surface of the Earth, drowning the good agricultural soil below.

Figure 4-3 Olivine as found on Earth in solid crystal form

The mixture of water and olivine then started to harden, like cement. The hardening process released heat. The heat drove water in the mixture upward to form a fog or mist. It is very possible that the Rio Cuarto Comet caused the growing problem on day 3 of God's work. The timelines proposed by scientists fit the events of day 3.

The LORD's laws of nature eventually solved the problem of no rain. The melting glaciers, resulting in rising ocean levels, caused rain to fall. Rain only needed time, before it started to fall. But the rain

---

1 https://en. wikipedia. org/wiki/Olivine
2 https://en. wikipedia. org/wiki/Silicon_dioxide#Chemical_reactions
3 https://creationscience. com/onlinebook/Comets. html
4 http://www. nature. com/nature/journal/v490/n7418/full/nature11469. html
5 D. J. Eicher. *Comets - Visitors From Deeppace.* New York: Cambridge University Press, 2013.
6 B. L. de Vries. Comet-like mineralogy of olivine crystals in an, 2012.

did no good as long as the hard coating on the land sealed the rain from the land below it.

What about the problem of no one to work the soil? People already lived on the Earth. Why could God not use the existing inhabitants of the area to work the soil? As I discuss in the next section, the problem was so severe that the LORD formed a new being to do the work. Figure 4-3[1] shows the mineral olivine in its pure form.

God Solves the World's First Labour Problem

The previous section shows how God needed someone to work the soil, "and there was no one to work the ground."[2] How did God intend to promote agricultural plants? The answer lies in the working of the soil. God planned on using man on Earth to work the soil and grow agricultural plants. The hard surface covering the Earth presented a problem. Water was available for plant growth, but a hard surface covered the land. As ground covered with a sidewalk, plants could not grow. Someone must turn over the top layer of soil to expose the good ground below.

The original Bible does not use the word "work," but a word that means "cultivate." The Hebrew word for cultivate has a connotation of "to put into servitude." Rather than the word "cultivate" in the normal sense of the word, this word implies ploughing, like a farmer breaking new land never in production. Ploughing turns over the soil, exposing soil below the Earth's surface. As well, history shows that a change in agricultural methods was in order. The natural method of letting seeds fall on the Earth's surface was not ideal for maximizing growth.[3] Scratching the surface of the ground to cover seeds with soil started during this time. This practice made it easier for seeds to sprout and resulted in less seed loss to birds and rodents. Thus, God's plan to cultivate not only helped the Earth recover from

---

[1] https://en. wikipedia. org/wiki/Olivine- Olivine by Rob Lavinsky is licensed under Creative CommonsAttribution-Share Alike 3. 0 Unported.
[2] Genesis 2:5
[3] R. Chadwick. *First Civilizations*. Oakville, CT, USA: Equinox Publishing Ltd., 2005.

the hard layer covering the Earth, but it also provided a means of increasing plant growth, in the future.

To start the growth of plants, someone must expose the good soil below. Science tells us that humans existed on Earth. They had the intellect to build shelters for themselves. Even in the early settlements such as Tell Abu Hureyra, they practiced rudimentary harvesting of wild grains.[1] Yet God could find no one to work the soil. Perhaps existing humans on Earth were just too lazy to do hard work. Ploughing and turning over new unprepared soil for planting is not an easy job. I can remember stories about my uncles and grandfather coming to Canada to start new farms. They had horses and ploughs to break new land, but it still was a very toilsome and backbreaking job. On day 4 of God's work, there were no ploughs, making his job especially backbreaking. Perhaps existing humans thought that the increasing light on day 3 meant they could return to their previous hunter-gatherer existence. Whatever the reason, the LORD God intervened to solve the problem. He formed Adam.

My religious education taught me that God made man on day 6. However, as we see in chapter 2 of Genesis, God needed someone to work the Earth on day 3. The Bible allows us to calculate Adam's age. I will discuss Adam's formation date later; however, suffice it to say that the Bible definitively places Adam's formation on day 3, about 9,400 years ago.

> Then the LORD God formed a man from the dust of the ground and breathed into his nostrils the breath of life, and the man became a living being.[2]

Verse seven tells us that the LORD God "formed" rather than "created" man from the dust of the ground. The Hebrew word for the English word "form" means to mold or shape, as in making a

---

[1] J. Ridout-Sharpe. Changing lifestyles in the northern Levant: Late Epipalaeolithic and early Neolithic shells from Tell Abu Hureyra. Quaternary International, 2015.
[2] Genesis 2:7

vase from clay. The original Hebrew word for man means a person of low grade. It is curious how this relates to stone tablet writings of early Mesopotamians. Early Mesopotamians believed that the gods formed man from clay.[1]

The LORD breathed the breath of life into man. The original Hebrew word for "breathed" implies a "forceful breath." The original Hebrew word used for the English words "into his nostrils" implies a patient and accommodating fury or rage. An interpretation of verse 7 could be "The Lord moulded the low-grade man out of the dust of the Earth and forcefully breathed into him a spirit of anger or impatience."

Why did the breath of life that LORD breathed into Adam include a spirit of anger or rage? Perhaps it had something to do with the makeup of the surface of the soil, which covered the Earth. A hard layer covered the land and trees. Perhaps Adam needed a spirit of anger or rage to do battle with this bad surface. Existing man apparently did not have the inclination to do the work required. The spirit of rage probably made Adam different from existing humans.

At this point, I should point out that the above discussion does not negate the formation of Adam on day 6. The LORD God's day 3 work caused the formation of a physical man. On day 6, God confers his spirit upon the physical man.

Formation from Clay Does Not Debunk Evolution

As we see in the previous sections, God "formed" Adam from the dust of the Earth. Does a literal interpretation of this verse invalidate the theory of evolution? If God literally formed Adam from dust, then is the chain of evolution from previous Homo sapiens broken? I think not.

When I was a young boy, back in the mid–1950s, *Flash Gordon* was on TV. *Flash Gordon* was an imaginary adventure series where humans travelled to faraway places in space. At the time, travel in space was unheard of, and we regarded what we watched on TV as

---

[1] https://en. wikipedia. org/wiki/Creation_of_man_from_clay

pure fantasy. Then the Soviet Union launched Sputnik into space, and the space race was on. It was not long before men were walking on the moon. Now we are living in space and we send missions to faraway planets and beyond. Pure fantasy became reality and almost a common occurrence today.

Today we watch movies like *Jurassic Park* where people bring prehistoric animals to life. This fantasy involves using preserved DNA to bring dinosaur-like animals back to life. However, just as Flash Gordon was, the recreation of dinosaurs is a fantasy. Could we someday really bring back dinosaurs? I think the answer is yes. If man can achieve this feat, then how much easier would it be for God to carry out?

In Genesis, God forms man out of the dust of the earth. The earth supplies the basic raw materials to form man. However, more importantly, every living species needs a plan or blueprint to fit all this matter together. This is the role of DNA. This is a concept that only a short time ago would have seemed more like fantasy than reality. As the movie *Jurassic Park* illustrates, DNA from ancient organisms also comes from the dust of the Earth. When God formed Adam, he knew how to take the carbon and other raw materials from the dust of the Earth, to make a living species. Of course, anyone can take the same raw materials and combine them. The only problem is that a person would only end up with a hodgepodge of mud. God on the other hand knew about the DNA from previously created organisms; it was part of the LORD's creation. In forming man on the third day, God started with existing DNA. Perhaps God made some small changes to the existing DNA that he used for the newly formed man. Perhaps this made the new man more suitable for God's intentions at the time.

The story of the formation of Adam does not debunk the theory of evolution. The Bible tells us that the newly formed Adam was still a being of "low grade." As such, God seemingly formed him with improved qualities, but he probably still was not that much different from other humans of the time. Science tells us that the DNA of the descendants of Adam, meaning us, is nearly identical to the DNA of previous Homo sapiens that lived before Adam and even like the

DNA of monkeys. Thus God, in forming Adam, used the same blueprint for modern humans as he used for the very first humans on Earth. While physically separate from previous Homo sapiens, we and Adam still evolved from early Homo sapiens.

We can also interpret the forming of Adam from the dust of the Earth nonliterally. The LORD God tells Adam and Eve, "For dust you are and to dust you will return."[1] Obviously, the LORD God views man in his living state as dust. It is possible that in forming Adam from dust, the LORD God looked at existing Homo sapiens on Earth as dust. We can then apply a nonliteral interpretation to the text. With this interpretation, the LORD God formed Adam from other humans already on Earth.

The nonliteral interpretation fits the theory of evolution to a tee. Under this interpretation, Adam, with some help from the LORD God, evolved from earlier life just like any other plant or animal life since the beginning of time. I will leave it up to you to decide which interpretation is more valid.

The Garden of Eden Story Happened on Day 3 or 4

Adam did not start cultivating the soil immediately. The LORD God planted a garden and placed Adam in it. The LORD God planted fruit trees in the garden. "Now the Lord God had planted a garden in the east, in Eden; and there he put the man he had formed. The Lord God made all kinds of trees grow out of the ground—trees that were pleasing to the eye and good for food. In the middle of the garden were the tree of life and the tree of the knowledge of good and evil."[2]

---

[1] Genesis 3:19
[2] Genesis 2:7

Figure 4-4 - The Persian Gulf today

The location of the Garden of Eden is of prime importance in deciphering the Bible story. Modern Bibles do not translate the applicable Hebrew Scriptures properly, making this task difficult. Before we look at the Hebrew Scriptures in detail, we need to study the topography and ice age history of the Persian Gulf.[12]

Figure 4-4 shows the Persian Gulf as it exists today. The waters of the Persian Gulf stop at the southern coastline of Kuwait. This was not the case before the six days of God's work.

Figure 4-5 depicts the Persian Gulf fifteen thousand years ago. The sea level was low because the glaciers had not yet melted. The Persian Gulf was essentially dry, except for small lakes. The ocean started at the Straits of Hormuz. Dry land existed between the Strait of Hormuz and the present day northern coastline of the Persian Gulf. The Euphrates, Tigris, Gihon, and Pishon Rivers gathered into one river named the Ur-Schatt River. The Ur-Schatt River ran from the present day northern coastline of the Persian Gulf to the Straits of Hormuz.

---

[1] http://Earthscience. stackexchange. com/ questions/3147/extent-of-the-persian-gulf
[2] Generally Taken from Douglas J. Kennett and James P. Kennett - Early State Formation in Southern Mesopotamia: Sea Levels, Shorelines, and Climate Change - Journal of Island & Coastal Archaeology, 1:67–99, 2006

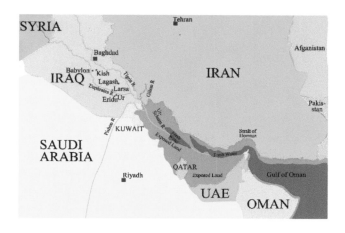

Figure 4-5 – The Persian Gulf on day 3

"Now the Lord God had planted a garden in the east, in Eden."[1] The original Bible text varies from modern translations. Instead of the English words "in the east, in Eden," the original Bible uses two Hebrew words: "Eden" and "eastward." "Eden" means the territory where Adam lived. "Eastward" means the direction "east" or "east of." It is important to note that the name "Eden" is separate from the word "garden." The original wording places the garden east of Eden, not in Eden.

The exact location of Eden is uncertain, but the Bible gives us an idea of where the location is. "A river watering the garden flowed from Eden; from there it was separated into four headwaters."[2] Modern translations are very misleading. The word "headwaters" does not appear in the original Hebrew text. The original Bible text reads similar to "and a river appeared beside Eden to water the garden, where it divided and turned into four main [or major] rivers." Thus, the river did not flow through Eden, but by or beside Eden. However, the river did flow through the garden, which was east of Eden.

---

[1] Genesis 2:8
[2] Genesis 2:10

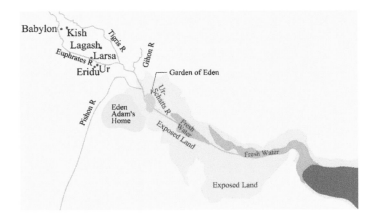

Figure 4-6 – The location of Eden and the Garden of Eden

The Bible goes on to name the rivers in the order that they join the main Ur-Schatt River; the Pishon, the Gihon, the Tigris, and the Euphrates. We know where the Tigris, Euphrates, and Gihon Rivers join today. However, the Pishon River dried up and we cannot see it today. The Bible tells us that the Pishon River was the first river to join. This means that the Pishon is the first river of the four or the southmost river of the four rivers that joined to form the Ur-Schatt River.

The Bible tells us that the Pishon River started in the mountains on the west side of present-day Saudi Arabia. "The name of the first is the Pishon; it winds through the entire land of Havilah, where there is gold. [The gold of that land is good; aromatic resin and onyx are also there.]"[1] The reference to gold, resin, and onyx ties Havilah to the Saudi western mountains. The Pishon River is what scientists call the Wadi Al-Rummah[2] River today. Scientists traced the origin of the river to the Al-Abyad Mountain near Medina, Saudi Arabia. Again, the Bible agrees with modern science.

Scientists say that the old basin of the Wadi Al-Rummah or Pishon River forms the west boundary of present-day Kuwait. The

---

[1] Genesis 2:11–12
[2] https://en. wikipedia. org/wiki/Wadi_al-Rummah

ancient Pishon River joined the river system south of where the present river system reaches the Persian Gulf. This location is under the waters of the Persian Gulf today. The Garden of Eden was south of this point. Thus, the location of the Garden of Eden is under the waters of the Persian Gulf today, as shown in figure 4-6. Today deep layers of coastal sediment cover the location of Eden. Scientists advise that flow from the Mesopotamian mainland, to the Ur-Schatt River deposited a thick sedimentary layer over the area.[1] Thus, in addition to being under water, the Garden of Eden is also under a thick layer of sediment.

As time went on, the sea level rose due to melting continental ice fields. This flooded previously dry land north of the Straits of Hormuz. The Ur-Schatt River became shorter. By 10,500 years ago, during day 3 of God's work, the ocean had not flooded the area of the Garden of Eden. However, after this time, the waters of the Persian Gulf advanced to cover the Garden of Eden. This means that the Garden of Eden story could not take place on day 6 of God's work. The ocean did flood the area of the Garden of Eden on day 6. The Garden of Eden story could only happen on days 3 or 4, when the area of Eden was still dry land.

Although there is very little archaeological record of early inhabitants in Eden, scientists suspect that between twelve thousand and eight thousand years ago, people lived along the Ur-Schatt River.[2] As well, they advise that archaeological evidence suggests a pre-Ubaid presence along the northeast coast of the Persian Gulf. Evidence of occupation exists along now dried out inland lakes north of the Persian Gulf. Scientists reported small stone structures at various coastal locations, such as Kuwait.

---

[1] D. J. Kennett. Early State Formation in Southern Mesopotamia: Sea Levels, Shorelines, and Climate Change - Journal of Island & Coastal Archaeology, 1:67–99, 2006.

[2] D. J. Kennett. Early State Formation in Southern Mesopotamia: Sea Levels, Shorelines, and Climate Change - Journal of Island & Coastal Archaeology, 1:67–99, 2006.

Animals and Adam's Helper

"The Lord God said, 'It is not good for the man to be alone. I will make a helper suitable for him.'"[1] Modern translations of the Bible lead us to assume that God initially intended to make a human partner for Adam. This may not be the case. The original Hebrew Bible does not use the English word "make" in the contemporary sense of the word. The Hebrew version of the word means "to accomplish, select, appoint or confer."

Modern translations of the Bible go on to say, "Now the LORD God had formed out of the ground all the wild animals and all the birds in the sky."[2] The original Bible text does not include the word "now" at the beginning of the verse. It simply starts the verse with the Hebrew word for the English word "formed," which means to mould or form as a potter forms a vase out of clay.

The original Hebrew text does not use the words "wild animals." Instead, the original Bible uses the phrase "beast of the field." The Hebrew definition of the word "beast" is any animal that has the spirit of life in it. The Hebrew word for the English word "field" means a field or land in the country intended for agriculture. Animals that dwell on agricultural land are domestic animals. The version of the Hebrew word for "field" suggests that the land needed some attention, and so we can deduce that the time is early and the land had not completely recovered from the catastrophes prior to God's work. The fields are still suffering from the ravages of the previous celestial bombardments. The Hebrew word for "birds" means any bird in general. Science tells us that animals and birds were in existence hundreds of thousands of years before the time of Genesis. The Bible tells us that the LORD God's concern in this case was with evolving domestic animals.

As with the creation of Adam, we can interpret the term "out of the ground" literally or not. The LORD God could have literally formed the animals from the dust of the Earth and used variations

---

[1] Genesis 2:18
[2] Genesis 2:19

of preexisting DNA to make these animals. Alternately, the LORD God could have modified the DNA of existing animals to evolve domestic animals.

Scientists[1] tell us that rudimentary domestication started approximately nine thousand to ten thousand years ago. In 7000 B.C., small settlements started. Rudimentary mixed farms provided plant crops and animals for food. A few hundred years later, humans were exploiting their secondary products. Soon animals provided secondary food, such as milk and cheese. Some animals provided means of locomotion by pulling carts. As well, hunting of wild animals often provided food.

The date of the above events is a few hundred years before the date that the LORD God formed Adam out of the dust of the Earth. It appears that the LORD God started domestication of animals before he created Adam. Perhaps the LORD God tried to use existing Homo sapiens in a failed attempt to advance domestication of animals.

Contemporary religions teach that God created fish and birds on day 5 and animals, including man, on day 6. As discussed above, this is not the case. We will see in subsequent sections that God merely decreed that domestic birds and animal increase copiously on days 5 and 6.

Modern translations of the Bible go on to say, "He brought them to the man to see what he would name them; and whatever the man called each living creature, that was its name."[2] Taken in context with previous phrases, we can deduce that Adam named the domestic animals that the LORD God formed. The Bible goes on to tell of how Adam named the animals: "So the man gave names to all the livestock, the birds in the sky, and all the wild animals, but for Adam no suitable helper was found."[3] Again, the original Hebrew Bible does not use the term "wild animals" but "beasts of the field."

---

[1] Brill. Handbook of Oriental Studies - Section One - The Near and Middle East – A History of the Animal World. Netherlands: Koninklijke Brill j\V, Leiden, The Netherlands, 2001.
[2] Genesis 2:19
[3] Genesis 2:20

The fact that the process of naming animals found no suitable helper for Adam tells us that God intended to appoint one of the animals as Adam's helper. The Hebrew word for the English word "suitable" means "around," "above/beside," "detached" and has a connotation of a complementing mate. Obviously, God intended Adam's helper to be more than a subservient assistant. God intended Adam's helper as a counterweight and advocate to Adam.

Eve, Fall from Grace, Adam Starts a Family

The LORD God formed domestic animals, but Adam still has no suitable helper. To solve this problem, God made Eve from a rib extracted from Adam. "Then the Lord God made a woman from the rib he had taken out of the man, and he brought her to the man."[1] It is interesting that the first chapter of Genesis talks about God only. The first chapter does not talk about forming new life and describes processes happening on their own through the laws of nature. The above verse in chapter 2 talks about forming new life. In this case the LORD God carries out the action. Obviously God does not form new life on his own; new life is formed only with the help of the LORD.

Adam seems very happy with Eve and sees her as an extension of himself: "The man said, 'This is now bone of my bones and flesh of my flesh; she shall be called "woman," for she was taken out of man.'"[2] Adam was ecstatic over the addition of his new mate. But this joy did not last long.

When the LORD God placed Adam into the Garden of Eden, he forbade Adam from eating from one tree: "And the LORD God commanded the man, 'You are free to eat from any tree in the garden; but you must not eat from the tree of the knowledge of good and evil, for when you eat from it you will certainly die.'"[3]

---

[1] Genesis 2:22
[2] Genesis 2:23
[3] Genesis 2:16–17

The story of Eve continues in Genesis 3. The infamous serpent apparently noticed some aspect about Eve's makeup that he could take advantage of: "Now the serpent was craftier than any of the wild animals."[1] The original Bible text does not use the term "wild" when referring to the serpent. The original Bible text uses the term "of the field" instead of "wild beasts," which indicates that the serpent is likely a domestic animal or some rodent that hangs around near domestic animals. The serpent tricks Eve into eating from the forbidden tree. God responds by expelling man from the Garden of Eden: "So God banished man from the Garden of Eden; So the Lord God banished him from the Garden of Eden to work the ground from which he had been taken."[2]

Adam and his wife had two children, Cain and Abel. Abel was a shepherd while Cain was a farmer: "Now Abel kept flocks, and Cain worked the soil."[3] The original Bible text uses a Hebrew word for "flocks" that means sheep or goats. The LORD favoured Abel's animal offerings more than Cain's grain offerings, prompting much envy on Cain's part. Cain kills Abel. As a result, God drives Cain from his presence.

Cain Leaves and Interacts with Existing Humans on Earth

I discussed how Homo sapiens existed before the six days of God's work. The Bible reflects what scientists tell us. Cain is apparently terrified; "Cain said to the LORD, 'My punishment is more than I can bear. Today you are driving me from the land, and I will be hidden from your presence; I will be a restless wanderer on the Earth, and whoever finds me will kill me.'"[4] Who were the people he feared? They had to be existing inhabitants of the Earth, alive prior to God's six days of work.

---

[1] Genesis 3:1
[2] Genesis 3:23
[3] Genesis 4:2
[4] Genesis 4:13–12

But the LORD put a mark on Cain that told others not to harm Cain. This seems to indicate that the LORD had some interaction with the preexisting populations on Earth. If the LORD's mark prevented them from following their feelings to kill Cain, they must have had some prior experience with the LORD, which made them fear acting on their desires and impulses.

Cain leaves the LORD's presence and travels to the land of Nod: "So Cain went out from the Lord's presence and lived in the land of Nod, east of Eden." [1]Cain's wife bore him children after he left for Nod: "Cain made love to his wife, and she became pregnant and gave birth to Enoch."[2] The Bible mentions no descendants of Adam during this period, except Cain and Abel. Cain had children after leaving God's care. It follows that Cain's wife was from existing inhabitants, not from the line of Adam.

If Cain married other inhabitants of the area, he started a separate species of man. Cain descended from Adam, whom the LORD formed out of the dust of the ground. Existing inhabitants of the area were not of this lineage. Cain's children then likely did not have all the qualities that Cain inherited.

Science[3] tells us that by 8,500 years ago, the Persian Gulf started to cover the area of the Garden of Eden. The LORD obviously expelled Adam and Eve from the garden before this time, at the end of day 4 or beginning of day 5. Cain's expulsion likely occurred a few years later, during the same general period.

Archaeologists tell us that various cultures started about eight thousand years ago. Karen Rhea Nemet-Nejat[4] identifies three cultures (the Hassuna, Samarra, and Halif) that evolved in northern Mesopotamia. The Hassuna period marked the start of agricultural and basic cloth products. A system of seals appears to indicate the

---

[1] Genesis 4:16
[2] Genesis 4:17
[3] D. J. Kennett. Early State Formation in Southern Mesopotamia: Sea Levels, Shorelines, and Climate Change - Journal of Island & Coastal Archaeology, 1:67–99, 2006.
[4] K. R. Nemet-Nejat. *Daily Life in Ancient Mesopotamia*. Westport, CT, USA: Greenwod Press, 1998.

identification of property ownership. The Samarra period witnessed the introduction of higher quality pottery. The Halif period introduced the tournette, a slowly turning wheel, presumably for pottery work. The Halif period saw the start of cobbled streets and workshop buildings, indicating the start of craft specialization. Buildings had many of the features described for Neveli Cori buildings. These cultures likely included residents from the previously abandoned settlement of Nevali Cori. These cultures lasted until about eight thousand years ago. Cultures of this era appeared fairly socially oriented. The Halaf society apparently was socialistic, involving central storage and sharing of goods. The Halaf culture shared this system with others.[1]

Scientists tell us the Ubaid culture started 8,500 years ago and lasted until 6,200 years ago.[2] The Ubaidians invented an axle and bearings, to which they added the Halif tournet, resulting in the first potter's wheel. The Ubaid culture started out as small farming villages in the south. Villages soon grew into large population centers. Larger cities evolved covering larger areas such as Eirdu at twelve hectares, Ur at ten hectares, and Uruk at seventy-five hectares. Large carefully designed temples appeared.[3]

Both the LORD's expulsions and movement of the Persian Gulf coastline northward caused Adam, Eve, Cain, and Cain's descendants to move. They undoubtedly started to interact with other humans in the area during this time. Science (Frangipane 2007) credits the Ubaid period as a time of increasing civilization. As social structure increased, so did canals and larger, more sophisticated buildings. It is no accident that civilization increased during this time. God forced Adam to leave the Garden of Eden, where he interacted with other residents of the area. As well, Cain started many generations of descendants, who also interacted with existing populations. These new peoples were from the root of Adam, a man specially created by

---

[1] M. Frangipane. Different types of egalitarian societies and the. *World Archaeology*, 2007.
[2] L. E. A. Khalidi. The growth of early social networks: Newgeochemical results of obsidian. *Journal of Archaeological Science: Reports*, 2016.
[3] K. R. Nemet-Nejat. *Daily Life in Ancient Mesopotamia*. Westport, CT, USA: Greenwod Press, 1998.

the LORD God to carry out his plan to promote domestic plants and animals on Earth. Their qualities obviously improved the fortunes of not only themselves but also existing residents of the area.

Science Enhances the Bible

In this chapter, we see how science makes the Bible story come alive. The formless and empty conditions cited in Genesis take on new meaning when we see the effect they had on existing humans at the time. As well, we debunked the myth that Adam was the first man on Earth.

We located the Garden of Eden. This is a good demonstration of how modern Bible translations present a misleading narrative. Then science came to our aid to show how the Garden of Eden story could only have happened on day 4 or 5 of God's work, not on day 6.

Looking at the history of Mesopotamia alongside the Bible, we find that the Bible is not an irrelevant book not in sync with the world, as it was at the time. To the contrary, we see how the Bible story fits with science and the development of early civilization of the world. The Bible story shows how the LORD God's intervention sought to rescue man from the previous devastating events on Earth. As we will see in chapter five, the LORD God established, through Adam and Eve, a growing civilization on Earth that improved the lives of many human beings.

5

Civilization Grows: Adam Becomes King

The interaction of Adam, Eve, Cain, and Cain's descendants with existing populations had a monumental impact on society. Populations increased, civilized life expanded, and life advanced to a new level of commerce, trade and know-how. This along with changes in climate soon put pressure on society.

My religious education taught me that God created fish, birds, and animals on days 5 and 6 of his work. As we learned in chapter 4, bird, fish, and animal life existed long before the time of Genesis. We also learned that the LORD expanded domestic plants on day 3. We will see in this chapter that God again expands birds, some types of fish, and domestic animals on days 5 and 6. This was in response to the increasing needs of a quickly developing society.

We will see in this chapter how God decides to hand over dominion of the Earth to Adam and Eve. Contrary to contemporary beliefs, God did not create Adam on day 6. He added his spirit to Adam and Eve on day 6, giving them the proper qualities for their new rule. We will see how science shows Adam was the first Mesopotamian king and how his kingship ushered in a new culture, the Sumerian culture.

Just as today, flaws burdened human nature. Man soon became corrupt and greedy. This eventually led to the destruction of mankind.

Cain: Rogue or Nation Builder?

Should we consider Cain a rogue or nation builder or both? The cultures of the day started various developments, such as rudimentary irrigation, so Cain possibly had some effect in moving society forward.

The Persian Gulf started to move northward after God expelled Cain from his care. Existing peoples in southern Mesopotamia had to move north. The declining land base crowded people closer together and caused competition for scarce resources. Cain and Adam were undoubtedly part of this process.

God expelled Cain from his care, approximately 8,600 years ago. Cain evidently was a very industrious person: "Cain was then building a city, and he named it after his son Enoch."[1] There is no direct evidence regarding the location of Cain's city. Perhaps some of the later settlements of early inhabitants discussed in the previous section are the remnants of settlements started by Cain and his descendants.

One of the earliest known settlements in the area is Chogha Mish.[2] This settlement dates to 8,800 years ago. This fits a timeline where Cain or his son could have built cities. This is right after the time that Cain left for Nod. Could Chogha Mish be the city Cain named after his son Enoch?

Some of Cain's descendants achieved notoriety, according to the Bible: "His brother's name was Jubal; he was the father of all who play stringed instruments and pipes."[3] One of Cain's descendants, Tubal-Cain, apparently worked with bronze and iron; "Zillah also had a son, Tubal-Cain, who forged all kinds of tools out of bronze and iron."[4] The Mesopotamian Bronze Age[5] started 4,900 years ago.

The Ubaid culture formed during the time of Adam, Eve, and Cain. Cain married after leaving God's care. Very likely Cain mar-

---

[1] Genesis 4:17
[2] https://en. wikipedia. org/wiki/Chogha_Mish
[3] Genesis 4:21
[4] Genesis 4:22
[5] https://en. wikipedia. org/wiki/Bronze_Age

ried an Ubaid descendent. Cain likely worked with the Ubaid people. He built a city. The inhabitants of the cities could only come from inhabitants already in the region. But perhaps Cain brought with him some negative traits that spread throughout the surrounding cultures. Science[1] tells us that society changed during this time. A more class-oriented society evolved. Some households started to become more prevalent over others.

The Moving Persian Gulf and Crowding

Science[2] tells us that rising ocean levels resulted in increased rainfall ten thousand years ago, or near the end of day 3 of God's work. This ended the semiarid conditions of the region, changing Mesopotamia into a lush, green land.

As with any large rainfall today, inland flooding occurs. River water, trying to reach the oceans and surface water and the rivers, backs up and floods inland regions. The same thing happened in Southern Mesopotamia. Now, not only the rising ocean levels flooded previously dry land. Fresh water trying to reach the oceans backed up and flooded land north of where the Persian Gulf ended. Flooding at present-day Kuwait started approximately 8,500 years ago. Adam had to move north. Water flooded the area of Eden. As well, the Ubaid culture had to move north.

---

[1] M. Frangipane. Different types of egalitarian societies and the. *World Archaeology*. 2007.
[2] D. J. Kennett. Early State Formation in Southern Mesopotamia: Sea Levels, Shorelines, and Climate Change - Journal of Island & Coastal Archaeology, 1:67–99, 2006.

Figure 5-1 shows the Persian Gulf region about six thousand years ago. Inland flooding extended as far north as the ancient cities of Ur and Eridu.

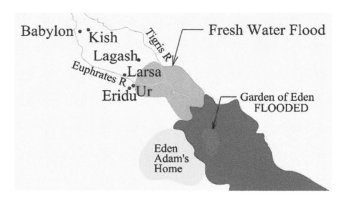

Figure 5-1 - Inland flooding as far north as Eridu

Scientists[1] believe that small villages were more common during the earliest phases of the Ubaid culture. The largest villages had less than one thousand inhabitants. Widely dispersed and un-connected villages were common. Scientific investigation shows evidence of rudimentary irrigation. Settlement followed the boundary between salt and fresh water as ocean levels rose and rain flooded inland areas.

As land became flooded and people moved north, competition for land began. The available resources land decreased in the face of an increasing population. This perhaps explains why God decided to provide more resources for people on day 5 of his work. Day 5 appears as a time of relative plenty, with all inhabitants living quite well. Some stress on local resources may have begun to appear though. Some scientists[2] feel that increasing population caused competition, resulting in a movement elsewhere to access available resources.

---

[1] D. J. Kennett. Early State Formation in Southern Mesopotamia: Sea Levels, Shorelines, and Climate Change - Journal of Island & Coastal Archaeology, 1:67–99, 2006.

[2] D. J. Kennett. Early State Formation in Southern Mesopotamia: Sea Levels, Shorelines, and Climate Change - Journal of Island & Coastal Archaeology,

Day 5: God's Response to Increasing Needs

As we discussed earlier, God formed from the dust of the Earth some kinds of birds and domestic animals before day 5, possibly as early as the end of day 3. Now, after a thousand years or more, God is ready to expand birds and fish vastly in response to the increasing needs of man.

The Bible uses a method in the next two verses to describe the expansion of fish and birds. The first verse introduces the general subject of fish and bird species. The next verse defines in detail the types of creatures involved. The first verse provides a general description: "And God said, 'Let the water teem with living creatures, and let birds fly above the Earth across the vault of the sky.'"[1] Again the Hebrew wording means to assign or declare and makes no mention of making or forming anything. On day 5, God expands what is already in existence.

The Hebrew word used for the English word "teem" means to sexually reproduce plentifully or copiously. The word implies a creature that wiggles or wriggles. It also implies creatures that teem or throng in large numbers. The Hebrew word for "living creature" indicates moving creatures that travel together. The last part of this general verse refers to birds. The Hebrew prefix for "and" appears as part of Hebrew word for "birds," making the modern translation correct.

The next verse narrows the definition of the creatures generally described in the first verse. "So God created the great creatures of the sea and every living thing with which the water teems and that moves about in it, according to their kinds, and every winged bird according to its kind."[2] As discussed earlier, the Hebrew word for "create" means to select or choose. The original Hebrew word for "and" means "more specifically." The original verse reads more like "God selected the great creatures of the sea, more specifically every living thing with which the water copiously reproduces and slithers around." The Hebrew word

---

1:67–99, 2006.
[1] Genesis 1:20
[2] Genesis 1:21

for word "teems" means to sexually reproduce plentifully or copiously. The Hebrew word for the English word "moves" has special meaning. The word means to scramble or proceed with a quick slithering motion or a series of small rapid steps. This limits the definition of "every living thing" to fish with fins that glide through the water. The definition also includes animals such as seals that crawl with a series of quick steps. This excludes other things such as crab, lobsters and snails. The more specifically selected creatures serve as food for man. Interestingly, they also serve as food for larger sea creatures.

On day 5, God commands aquatic species and birds to multiply fruitfully: "God blessed them and said, 'Be fruitful and increase in number and fill the water in the seas, and let the birds increase on the Earth.'" [1]

The Hebrew word for "blessed" means to plentifully or lavishly endorse or sanction. The Hebrew word for "increase" means to multiply lavishly. God eased the pressing needs of Mesopotamians by increasing resources. If we look at Earth today, we see the power of God's blessing. Modern man enjoys an abundance of blessings, as the availability of fish and poultry products demonstrates today.

Day 6: More Resources for a Growing Population

The north shoreline of the Persian Gulf started a rapid advance northward, between 7,400 and 6,700 years ago. By the end of the period, the north coastline of the Persian Gulf extended northward to a city named Eridu.[2] Scientists believe that communities increased in size to as large a two thousand to three thousand people. The intensity of rainfall probably reached a maximum during this time. Then rainfall started to decline, and temperatures started to fall. The advance of the Persian Gulf north slowed down.

Around 6,800 to 5,500 years ago, annual precipitation dropped by fifty millimeters and temperatures fell by 2°C in July and by 1°C

---

[1] Genesis 1:22
[2] https://en. wikipedia. org/wiki/Eridu

for January. This resulted in more intense farming and a weakening in some plant species.¹ (Springer 2007)

There is evidence of the start of trade during this period. Along with trade came an increase in competition for land and resources. Man became more competitive and less concerned about his neighbors.²

On day 6, God comes to the rescue with more resources, in the form of increased numbers of domestic animals: "And God said, 'Let the land produce living creatures according to their kinds: the livestock, the creatures that move along the ground, and the wild animals, each according to its kind.'"³ As with birds and fish, the word "said" means to assign or declare, and the original Hebrew Bible text tells us that the creatures will emerge or appear on good agricultural land. As on day 5, the Bible tells us not of an act of making something, but the act of expanding something already in existence. The original Hebrew Bible then lists specific animals involved. The first is livestock. Modern translations characterize the second kind of animals as "the creatures that move along the ground," such as reptiles like frogs, snakes, and lizards. The Hebrew word for the third type of animal could include wild animals, but the original Hebrew Bible does not include the word "wild." The original Bible uses two Hebrew words that mean "and beast" and "of the Earth." Earth means in geographical sense, and so the third type of animal could include both domestic and untamed animals.

God then narrows the kind of animals involved. "God made the wild animals according to their kinds, the livestock according to their kinds, and all the creatures that move along the ground according to their kinds."⁴ The original Hebrew word for the word "made" means to accomplish, carry forward, or turn into, similar to a farmer

---

[1] Springer. *The Black Sea Flood Question:*. Dordrecht, The Netherlands, The Netherlands: Springer, 2007.
[2] D. J. Kennett. Early State Formation in Southern Mesopotamia: Sea Levels, Shorelines, and Climate Change - Journal of Island & Coastal Archaeology, 1:67–99, 2006.
[3] Genesis 1:24
[4] Genesis 1:25

selectively breeding certain types of qualities into a line of animals. The original Hebrew Bible does not include the word "wild." Instead the Bible uses two words that mean "animals" and "of the earth." The word for earth means land, such as farmland, narrowing the selection to domestic animals. The Bible then limits the types of domestic animals to two types, livestock, and reptiles.

On day 6, God brings forth all animals on the Earth with a special emphasis on animals that live close to man on cultivated land. Livestock provides food for man. The emphasis on reptiles is interesting. Perhaps God expanded reptiles to keep control of insects, such as flies, that tend to increase rapidly around livestock.

Day 6: Handing over Dominion

But life was not great for all people. Greed and selfishness emerged. Society changed to a more competitive system, which allowed certain groups to prosper at the expense of others. The upper echelon of society tended to widen the gap between rich and poor. This destabilized society, which became more dynamic and ready for change.[1]

At this point, God decided to hand over control of things on Earth to Adam and Eve. As I discussed above, God's work on day 5 and the beginning of day 6 was likely an effort to supply the needs of the inhabitants of the area and to reverse the increasing competition and developing bad behavior of the time. Handing over control of the Earth to Adam and Eve very likely was an attempt to change the increasingly competitive and uncaring nature of the population.

Nevertheless, as discussed before, in chapter 2 of Genesis, Adam and Eve were still beings of low grade. Although set aside, from other Homo sapiens, they still were not that much more superior to them. God needed someone with superior qualities to take dominion of the Earth. God made Adam and Eve into superior beings by imparting spiritual qualities to them. Homo sapiens alive at the time obviously

---

[1] M. Frangipane. Different types of egalitarian societies and the. *World Archaeology*, 2007.

did not have the qualities to take on this task. The prospective candidates required more God-like qualities. However, God is a spirit. A spirit is invisible. So how did God bring about a man with enhanced spiritual qualities? "Then God said, 'Let us make mankind in our image, in our likeness, so that they may rule over the fish in the sea and the birds in the sky, over the livestock and all the wild animals, and over all the creatures that move along the ground.'"[1] The Hebrew word used for "said" means to assign, endorse or certify.

The first part of the verse, "Let us make mankind in our image, in our likeness," gives us an insight regarding the process God used to accomplish the task. The Hebrew word used for the English words "Let us make" means to achieve, take forward, assign, select, turn into, or confer.

In this verse, the Bible makes a very important distinction. I used the term "Adam" previously in this book to maintain clarity. Before day 6, the Bible used versions of the Hebrew word "man," not "Adam." This verse uses a new version of the Hebrew word used for the English word "mankind" that means "Adam." Previously, when God created man out the dust of the Earth, the Bible used a word for the English word "mankind," which means a being of low grade or low quality. When God confers his spirit on what previously was merely a "man," the man's name changed to "Adam."

Modern translations vary from the original Hebrew Bible regarding the phrase "in our image, in our likeness." The original Hebrew text uses only two words for the phrase "in our image, in our likeness." The Hebrew word for "in our image" means a ghostly apparition or a symbolic spiritual representation. The Hebrew word for "in our likeness" means a similarity, sameness, or representation. The next verse shows how the images conferred are of a special quality, conferred nowhere else in the Bible.

"So God created mankind in his own image, in the image of God he created them; male and female he created them."[2] The sentence structure of this verse in the original Bible is different from

---

[1] Genesis 1:26
[2] Genesis 1:27

that in modern translations. The English words "So God created mankind in his own image" stand alone, as a separate sentence. The word "created" means to "select" or "send out." The Hebrew word for "mankind" means a being of low stature or grade. Another translation for this sentence might be "God sent out the man of low grade with God's spiritual image."

The rest of the verse is a single sentence: "In the image of God he created them; male and female he created them." The first part of the verse refers to the addition of God's spirit to Adam and Eve. The verse ends with "male and female he created them." In this case, the Bible uses another Hebrew word "create," which means selection in a physical sense. The first part of the sentence speaks about a spiritual selection. The second part of the verse tells us that God selected Adam, the man of low grade formed by the LORD from the dust of the ground and Eve who the LORD made from Adam's rib. Thus, while Adam and Eve were still physical beings, like others on Earth, they now were also special. Unlike other humans on Earth, they had God's spirit.

"God blessed them and said to them, 'Be fruitful and increase in number; fill the Earth and subdue it. Rule over the fish in the sea and the birds in the sky and over every living creature that moves on the ground.'"[1]

The modern translation of the first part of the verse, "Be fruitful and increase in number; fill the Earth and subdue it," is not accurate. A more accurate translation of the first part of the verse is "be fruitful and multiply and replenish the Earth." The original Bible text uses a Hebrew for the English word for "replenish," which means to achieve or sanction. The word has connotations of "to be enclosed or hedged" or "to be concluded." The word for "Earth" means land. God is telling Adam and Eve to give special sanction to domestic land and animals.

Interestingly the phrase "every living thing that moves on the Earth" includes more than just animals. It includes other humans as well. The Hebrew word for "ground" means soil. Seemingly, God's

---

[1] Genesis 1:28

intention is to give Adam and Eve control over the food supply of the Earth as well as all humans who produce it.

God not only turned plants, animals, birds, and fish to Adam. God also turned over kingship of all humans over to Adam. This happened at the end of day six. Scientists tell us that the Ubaid period ended shortly after this time, ushering a period of unprecedented growth in populations, commerce and trade. It is not just a coincidence that the addition of God's spirit to Adam and Eve ushered in this increase in activity.

The Uruk Period and the Sumerian Culture

Scientists[1] (Sundsdal 2011) tell us the Uruk[2] period started approximately six thousand years ago and lasted until five thousand years ago. The period marked a time of accelerated innovation, urbanization, economic and social transformations, and advanced products such as pottery, architecture, art, and information recording devices, like the cylinder seal, spread and replaced local traditions. Commerce resulted in the trading of Uruk material that appeared in other regions/territories. This period saw the emergence of trade within Mesopotamia and nearby surrounding countries. Governmental structures became more intense and organized. The settlements demanded taxes on exported goods. Increasing technology such as seals and numerical tablets supported trade and taxation. The elite used seals, indicating increasing worries about the control of property and goods. Seals on containers and buildings restricted access to goods and raw materials to only the authorized. Southern Mesopotamian influence also spread throughout the region, making it the center of trade networks. Its administrative systems spread and adjacent territories adopted them. Adjacent regions accepted Mesopotamia as the world's reference point for economic matters and for knowledge and technology. Seals displayed Mesopotamian

---

[1] K. Sundsdal. *The Uruk Expansion: Culture Contact, Ideology and Norwegian Archaeological Review*, http://www. tandfonline. com/action/showCitFormats?-doi=10. 1080/00293652. 2011. 629812, 2011.

[2] https://en. wikipedia. org/wiki/Uruk_period

religion and myths on goods shipped throughout Mesopotamia and adjacent regions.

A strange new culture emerged shortly before this period as well—the Sumerian culture. Archaeologists have very little data to track the origins of the Sumerian culture. The earliest indications of the Sumerian culture start about 5,100 years ago.[1] Many scientists base the date 5,100 years ago on the advance of cuneiform, a system of writing first developed by the ancient Sumerians of Mesopotamia.[2] Although the writing system came into common use 5,100 years ago, its development started earlier in time. Other scientists feel that the Sumerian culture arrived six thousand years ago.

Sumerians appeared as the Persian Gulf receded, exposing a new southern Mesopotamian alluvial plain, which is now south Iraq.[3] During this time, rainfall decreased, and the fresh water lake that flooded the land as far inland as Eridu receded, exposing more land.

Scientists[4] put forth many unproven theories about the origins of the Sumerian culture. Most scientists seem to agree that the Sumerian language has no demonstrable genealogical relationship with other languages.[5] There is a simple answer to the question. It appears that God started the Sumerian culture as part of imparting his spirit to Adam on day 6. The date is six thousand years ago, during day 7. This is after God implanted his spirit in Adam. The arrival of the Sumerian culture right around the time of Adam's kingship is no accident. The kingship of Adam, with his newly acquired spirit of God, was the impetus that drove forward a huge advancement of society. The development of the cuneiform writing system, which was an order of magnitude advancement over existing writing systems, started 6,000 years ago and came into full power 5,100 years

---

[1] K. R. Nemet-Nejat. *Daily Life in Ancient Mesopotamia.* Westport, CT, USA: Greenwod Press, 1998.
[2] http://www. ancient. eu/cuneiform/
[3] J. Oates. *More Thoughts on the Ubaid Period.* Chicogo, Illinois, USA: The Oriental Institute, 2006.
[4] https://en. wikipedia. org/wiki/Sumer
[5] https://en. wikipedia. org/wiki/Language_isolate

ago. Figure 5-2[1] shows a tablet written in cuneiform. The predominant language spoken in southern Mesopotamia 5,000 years ago was Sumerian.[2]

Figure 5-2—Sumerian tablet showing a bill of sale for a slave

(Judaica Encyclopaedia 2007) Sumerians lived in towns and cities with narrow, unpaved, and unsanitary streets.[3] Families lived in multiroomed houses, made of mud and bricks. Less numerous and more elegant two-story houses existed for the elite. The population of towns and cities varied from ten thousand to fifty thousand. Residents of towns and cities regarded a large temple complex, as the center of their life. The population consisted of free citizens, serflike clients, and slaves. Some of the free citizens served as high temple officials. Together with rich landowners, this group formed an elite noble class. High officials, subject to a priest king, governed. The king had ultimate power, but he was responsible for the best interests of the people. The people considered the king an intermediary with their gods, of

---

[1] https://en. wikipedia. org/wiki/Sumer- Bill of Sale by Louvre Museum is licensed under Creative CommonsAttribution 2. 5 Generic
[2] K. R. Nemet-Nejat. *Daily Life in Ancient Mesopotamia*. Westport, CT, USA: Greenwod Press, 1998.
[3] *Judaica Encyclopaedia*. "Sumer, Sumerian," 2007.

whom there were many. The king's responsibilities included assurance of the prosperity and well-being of the land by maintaining the irrigation system, taking care of the temples and through the preservation of laws and justice. Most free citizens were farmers and fishermen, artisans and craftsmen, merchants, and scribes. Serfs were dependents of the temple, palace, and rich estates. The family formed the basic unit of society. Society was male dominated, but women also had rights.

Agriculture was the prime industry in Sumer; however, Sumerians carried on a manufacturing industry.[1] Some products showed signs of mass production, indicating factories developed as time progressed. The quality of products indicated highly skilled craftsmen. Goods appeared to be everyday articles intended for the masses as well as intricate and exotic articles for the elite in an increasingly stratified society. Some cottage-style operations existed as well as larger factory-type operations. The temple nobility ran the factories and paid craftsmen with land and food stuffs. Money was not in existence; however, the elite devised systems of valuation based on a commodity such as copper. Textile manufacture was the next most important industry after agriculture. Evidence suggests that women manufactured textiles in temple or palace workshops. Pottery manufacture was another major industry. Pottery manufacture took place in smaller workshops in various towns and cities. Metalworking on the other hand required valuable materials. Some ores seem to have come from the Iranian Plateau. Metal smiths were very skilled and worked mainly with lead, silver, gold and tin. Some experimentation with iron also took place. Stone work, as well as other manufacturing, also took place in Sumer. Trade flourished in Sumer. Many of the products manufactured in Sumer required the import of materials from foreign countries. Rulers sometimes married off their daughters to foreign rulers in a diplomatic effort to improve trading relations. Sumerian imports increased steadily throughout the period. The city state of Uruk apparently dominated as a center for world trade stretching to the Anatolian plateau in the north and Elam in the

---

[1] H. Crawford. *Sumer and the Sumerians*. Cambridge, UK: Cambridge University Press, 2004.

southeast and much farther as time went on. Trading colonies and stop-over settlements along trade routes developed. Water ways were the most popular means of shipping, however, some used donkey, cattle or human driven carts for overland hauling. Boat shipments, to and from centers south of present day Kuwait, also occurred on the Persian Gulf. Sumerian[1] civilization significantly affected inhabitants. Eventually, it seems like the Sumerian culture swallowed up the Ubaidian culture by 5,800 years ago.

The Sumerian King List: Adam the First King

Figure 5-3 - Sumerian King List

Ancient stone tablets known as the Sumerian King List, shown in figure 5-3[2], record the kings of Samaria, starting from the beginning of the Sumerian culture. Dates shown on these tablets used an ancient system of time units known as "sars" and "ners." Scientists have tried to calculate the duration of rule for kings listed in the tablets. For some reason, scientists assumed that a sar equals 3,600 years and a ner equals 600 years. This resulted in unrealistically long times, such as 28,800 years.

Scientists provided us with two markers in time, to evaluate the Samarian King List. Science[3,4] tells us that the city of Eridu,[5] the oldest city in the world, began approximately seven thousand years ago. As we will see below, this is around the time that Adam received kingship. As discussed

---

[1] https://en. wikipedia. org/wiki/Sumer
[2] https://en. wikipedia. org/wiki/Sumerian_King_List- Sumerian King List by Chinese Wikipedia projectis Public Domain
[3] P. Charvat. *Mesopotamia Before History.* New York, NY, USA: Routledge, 2002.
[4] R. Chadwick. *First Civilizations.* Oakville, CT, USA: Equinox Publishing Ltd., 2005.
[5] https://en. wikipedia. org/wiki/Eridu

in more detail in chapter 6, there is archaeological evidence of a great flood 4,900 years ago. Thus, a time span of 2,100 years elapsed between the founding of Eridu and the suspected evidence of the flood. The King List shows a total of sixty-six sars, plus six ners elapsing during this time. By this metric, one sar equals slightly less than 31.7 years. Thirty-one is close to the number of days in a month. One sar is also equivalent to approximately 365 months. 365 is close to the number of days in a year. Clearly, the ancient Sumerians had a system of time, which somehow relates to real time. I found that multiplying the number in years, quoted in contemporary literature, by 0.0087 (31 divided by 3600) gives timeline results that fit with other events in history. Current literature quotes the opening line in the Sumerian King List as "After the kingship descended from Heaven, the kingship was in Eridu. In Eridu, Alulim became king; he ruled for 28,800 years."[1]

Settlements at Eridu started seven thousand years ago. Astonishingly, this is right near the end of God's sixth day of work. It seems more than just a coincidence that Mesopotamian history shows the emergence of the first kingship at the same time God gave Adam and Eve dominion over the Earth. The spiritual remaking of Adam occurred 6,760 years ago, using Bible data. The date of Adam's spiritual birth fits the timeline that science tells us Eiridu came into existence. The ancient Mesopotamian figure "Alulim" came to Eiridu. Alulim and Adam likely are the same person. The Sumerian King List indicates that the first Sumerian kingship descended from heaven. Not only does the timeline fit but the King List indicates some form of heavenly intervention that reflects the Bible story of Adam receiving God's spirit.

History tells us that God had good reason to decree Adam as king. The period before and after Adam's kingship saw intense and rapid urbanization. The rapid development of civilization started to give way to some undesirable behaviour. Crowding, exacerbated by the northward movement of the Persian Gulf, led to intense competition for resources. As time progressed, the inhabitants became

---

[1] S. Langdon. *Oxford Editions of Cuneiform Texts.* OXFORD UNIVERSITY PRESS, 1923.

increasingly polarised. The concept of equality for all declined as society became more competitive. This resulted in social stratification as some households fell behind. Social mobility declined and migration between different social strata became more difficult. Temple leaders unequally distributed farmland and resources.

Science tells us that hereditary leaders ruled larger communities through central organizations.[1] This allowed ambitious individuals to control the best locations, where they gained competitive advantages. This created the basis for political groups at the end of the "Ubaid Period"

Some historians and scientists have speculated that an elite class of hereditary chieftains evolved. These chieftains likely greatly affected the administration of the temple shrines and their granaries. They likely were responsible for mediating intra-group conflict and maintaining social order.

Kingdoms Change: Resistance toward the LORD

Figure 5-4 - Map of southern Mesopotamia

---

[1] D. J. Kennett. Early State Formation in Southern Mesopotamia: Sea Levels, Shorelines, and Climate Change - Journal of Island & Coastal Archaeology, 1:67–99, 2006.

Figure 5-4[1] shows city states in Mesopotamia. The King List shows that kings came and went. It is likely that Adam's sons followed Adam as kings. Proponents quote the king list as saying that city-states "fell," at various times in history.

When Adam received kingship, the Persian Gulf was already quite some distance north, close to the city if Eridu. The King List shows the first kingship of Eridu falling 6,346 years before the year 2000 and the kingship moving to Badtibira. It is unclear if the move from Eridu to Badtibira had something to do with the advancing Persian Gulf and flooding of Eiridu.

Badtibira falls to Larag (Larsa), 5,495 years before the year 2000. Larag is south of Badtibira. This move seems to resist the northward movement trend. The end of the Ubaid period occurred during this time, ushering in the Uruk period[2] and increasing trade, governmental structures, and more complex, larger cities. A move from Larag to Zimbir (Sippar) 5,245 years before the year 2000 reestablished the northward movement. A long southward move from Zimbir (Sippar) to Shuruppak 5,062 years before the year 2000 supports the fact that the climate started to change to semiarid six thousand years ago. Perhaps the move to Shuruppak had something to do with declining agriculture in more northerly regions. As well, the receding waters of the Persian Gulf affected maritime commerce and trade. Declining river levels in northerly regions probably reduced access to shipping routes on the Persian Gulf to the south. Between five thousand and four thousand years ago, severe arid conditions returned.

Chapter 5 of Genesis ends with "At that time people began to call on the name of the LORD."[3] This occurred approximately 6,300 years ago, during day 7, when the LORD God rested. Contemporary translations give the impression that people were asking the LORD God for something or that they were worshipping the LORD God. A

---

[1] https://en. wikipedia. org/wiki/History_of_Sumer - Map with the locations of the main cities of Sumer and Elam by Ciudades_de_Sumeria (https://commons. wikimedia. org/wiki/File:Ciudades_de_Sumeria. svg) is licensed under Creative CommonsAttribution 3. 0 Unported
[2] https://en. wikipedia. org/wiki/Uruk_period
[3] Genesis 4:26

detailed review of the original Hebrew Bible indicates just the opposite. Instead the people began to defiantly defile or despoil the very famous name of the LORD. Apparently, the granting of dominion to Adam gave rise to some serious resistance. The residents apparently were aware of the LORD and recognized him as great. Now they rebelled against him. This happened between the end of the Ubaid period and the beginning of the Uruk period, approximately 6,300 years ago.

The Final Defilement

Adam and Eve probably had their hands full in ruling their new kingdom, but things got even worse. "When human beings began to increase in number on the Earth and daughters were born to them, the sons of God saw that the daughters of humans were beautiful, and they married any of them they chose."[1] The Bible uses a Hebrew word for "human beings" that means "man of low degree," or in other words, without God's spirit. The original Bible text uses no word that means "marry." The original Hebrew text uses an adjective in conjunction with the word "daughters" that implies "anyone" or "any sort."

There apparently were different types of humans on the face of the Earth. The majority were the descendants of ordinary men. These were descendants of people already were on Earth before the start of the six days of God's work. As well there were the sons of God. These were the descendants of Adam and Eve. They were of higher quality because they possessed God's spirit. Obviously, the physiology of the two lineages was very similar. The sons of God found the daughters of man attractive. They interacted sexually with them and probably had children with them.

"Then the LORD said, 'My Spirit will not contend with humans forever, for they are mortal; their days will be a hundred and twenty years.'"[2] The LORD shortens the life span of man as a

---

[1] Genesis 6:1–2
[2] Genesis 6:3

result. The original Bible does not use the word "humans." It uses a Hebrew word for "man" that refers to the general man of low degree. These are humans without God's spirit, as the words "for they are mortal" shows. The wording in the original Bible is different than modern translations and reads similar to "always also for he is flesh." The Hebrew word for "flesh" means something physical that is a carcass made of blubber, tissue, and a skeleton. The Hebrew word for "always" means for all time. Unlike Adam, who had God's spirit, ordinary humans possessed only a mortal body only. So God shortens the life of humans who do not have Adam's God-given spirit. This clears up one major criticism of skeptics. Critics (Bucaille n.d.) point out that while the LORD shortens the lives of humans to 120 years, several descendants of Noah lived much longer lives. The LORD shortened the lives of ordinary humans not from Adam's line. There is no contradiction in the Bible regarding the fact that Noah's descendants lived longer lives than 120 years.

God had a specific intent when he gave man resemblances of his spiritual images on day 6. He started a new type of human being with special qualities. Not only Adam and Eve receive these qualities; their descendants did also. The Bible's description of Seth, Adam and Eve's third son, demonstrates this. "When Adam had lived 130 years, he had a son in his own likeness, in his own image."[1] The Bible uses the same Hebrew word for "likeness," as used when God created Adam in God's likeness. The word for "image" in this case is similar, but implies a transmission from man to man instead of from God to man. Thus, Seth had God's image, just as Adam and Eve did. If Seth and his descendants married wives descending from the genealogy of Adam, their descendants also possessed God's image. When the sons of God interacted with ordinary females, their descendants evidently no longer possessed God's spirit. Much to the disappointment of God, the sons of God defiled and broke the line of spiritually enhanced humans formed by God.

---

[1] Genesis 5:3

Society Becomes More Corrupt

Predatory business practices evolved. Unfortunate people fell into debt and became distressed. Interest rates of 25 to 20 percent as well as failure to pay penalties decimated the poor, who then had to work for their creditors. The poor lived in continual fear of creditors and fear of reprisal from the palace. Workers were paid bare subsistence wages, while the upper class owned land to supplement their food supply.[1] It seems that the elite ruled and decided the fate of lower classes. Rations afforded often fell short of what the elite promised.

The Jemdet Nasr[2] period, between 5,100 and 4900 years ago, ended abruptly at the start of the flood. The Jemdet Nasr[3] period marked a period of increasing sophistication of commerce and trade.[4] Central administrations made society more controlled and organized. The emergence of highly elaborate, ornamental pottery suggests the emergence of an elite upper class. Society became more polarized and stratified. An elite class ruled the economy, at the expense of the lower classes. Distribution of goods was unequal and some parts of society suffered, while others thrived.

Conditions apparently reached a point of no return. The LORD lost patience with humanity: "The LORD saw how great the wickedness of the human race had become on the Earth, and that every inclination of the thoughts of the human heart was only evil all the time. The Lord regretted that he had made human beings on the Earth, and his heart was deeply troubled. Therefore, the Lord said, 'I will wipe from the face of the Earth the human race I have created—and with them the animals, the birds and the creatures that move along the ground—for I regret that I have made them.'" But Noah found favor in the eyes of the Lord."[5]

---

[1] K. R. Nemet-Nejat. *Daily Life in Ancient Mesopotamia*. Westport, CT, USA: Greenwod Press, 1998.
[2] https://en. wikipedia. org/wiki/Jemdet_Nasr_period
[3] https://en. wikipedia. org/wiki/Jemdet_Nasr_period
[4] https://en. wikipedia. org/wiki/History_of_Mesopotamia
[5] Genesis 6:5–8

6

The Flood and Restart of Life

The story of the great flood is one source of ammunition for detractors of the Bible. We address one major argument used in the first section of this chapter. As we discuss, detractors say that the numbers of animals entering the ark changes and the Bible contradicts itself. No contradiction exists when we take the nature of God discussed in chapter 1 into account.

The volume of water required to flood the Earth is a second major argument that the Bible detractors use to prove that the Bible is out of touch with science and reality. This chapter shows how recent scientific findings show that sufficient volumes of water exist below the earth to flood the Earth.

The Bible flood story stops approximately one year after the flood began, and many believe that this is the end of the flood. This is when the ark landed on the top of Mount Ararat, and the land around the ark dried. This was not the end of the flood. As we will see, it took many years for the waters to drop below the top of Mount Ararat to normal sea levels. We will see how science verifies the total length of the flood. I add my own theory about how Antarctica provides answers about the location of major return conduits to the deep.

It took hundreds of years for Noah's descendants to make it back to Mesopotamia, where they started from. Science provides a record of their arrival through archeological data that shows when ancient preflood cities were reoccupied.

This chapter ends the discussion about how the Bible fits with history. We look at how man soon again became corrupt. God selects Abraham to continue his line on Earth.

Flood Preparations: The Ark Passenger List Changes

When the LORD decides to end life on Earth, he turns detail arrangements over to God. God commands Noah to build an ark. The ark had three floors, each with an area of nearly fifty-three thousand square feet. The total area was nearly the area of an American football field, including end zones. The size of the ark certainly was of a size that could hold a huge number of animals and people. I estimate that the ark, if totally filled with adult cows, could house over three thousand cows along with enough hay to feed them for one year. The inside height above each floor of the ark was over fifty feet. This leaves ample room above the cow's heads to house birds and other small animals. The ark's large size certainly allowed it to carry many creatures.

The ark was the perfect design and of an advanced design that gave it stability and the ability to withstand storms. According to Hobrink, God did not save every type of animal on the face of the Earth. He saved classes of animals. He cites the example of dogs. God needed to take onto the ark only one male and one female dog. This pair then formed the basis for the evolution of other species on Earth, such as the various domestic and untamed dogs as well as wild animals such as wolves, coyotes, foxes, and jackals.[1]

"You are to bring into the ark two of all living creatures, male and female, to keep them alive with you."[2] God tells Noah which creatures will enter the ark. Noah builds the ark according to God's specifications. Detractors of the Bible use the flood story as a prime example of a major contradiction in the Bible. Note that in this verse, God gives instructions to keep one male and one female of every species alive.

---

[1] B. Hobrink (n. d.). *Science and the Bible.*
[2] Genesis 6:19

As discussed below, the number of animal changes. Some authors feel these changes are mistakes in the Bible.[1] Dr. Bucaille cites these verses of the Bible as blatant contradictions. He again blames the Yahvist and Sacardotal versions of the Bible as the source of error, where parts of the Bible refer to the name LORD and others to the name God as the same and only higher power. As we discussed in chapter 2, the LORD and God are different entities. This largely solves the apparent contradiction.

It is important to note that the LORD, not God changed the numbers of animals to save on the ark. Initially God instructs Noah to bring into the ark "two of all living creatures, male and female". This definition, as well as animals, includes other humans on Earth.

Then the LORD gets directly involved and changes the numbers of animals: "Take with you seven pairs of every kind of clean animal, a male and its mate, and one pair of every kind of unclean animal, a male and its mate."[2] The LORD changes the things to save on the ark to nonhumans only. Obviously, the LORD wanted to start clean. He decided humans to save included Noah and his family only; they possessed God's spirit, making them acceptable to the LORD. The LORD effectively overruled God.

The change in passenger list probably took other things into account as well. The numbers of clean animals and birds increased to seven pairs. The reason for this might be to have clean animals on board that Noah could use for food. Perhaps milk from cows and eggs from chickens provided food for Adam and his family. The ark floated on the flooded Earth for nearly a year. Some of the animals could have bred and given birth to offspring during this time. Perhaps the additional animals were a source of food for Adam and his family and indeed for other meat-eating animals on the ark.

---

[1] D. M. Bucaille (n. d.). The Bible The Qur'an and Science.
[2] Genesis 7:2

The Flood Comes

The LORD gives Noah a preview of the flood to come: "Seven days from now I will send rain on the Earth for forty days and forty nights, and I will wipe from the face of the Earth every living creature I have made."[1] Notice that rain is the action that will wipe all life from the face of the Earth.

Noah and his family enter the ark and then "pairs of clean and unclean animals, of birds and of all creatures that move along the ground, male and female, came to Noah and entered the ark, as God had commanded Noah."[2] It is interesting that the animals approached Noah on their own. Noah did not have to round them up. The last sentence is also interesting: "Then the Lord shut him in."[3] Many naysayers probably ridiculed Noah when he was building the ark. They likely became terrified when the waters started to rise and likely tried to enter the ark. Possibly the LORD shut in the ark to prevent additional passengers from embarking.

Several Mesopotamian stone tablet writings tell of a great flood as well. The Tale of Ziusudra[4] is one such story. The gods had decided to send a flood to destroy humankind. The god Enki (lord of the underworld sea of fresh water and Sumerian equivalent of Babylonian god Ea) warns Ziusudra, the ruler of Shuruppak, to build a large boat. Damage to the tablet obscures the writing; the next surviving part of the tablet describes a flood.

The Samarian King List lists some eight kings[5] before the great flood. The king list has the following entry for the flood "Then the flood swept over."

The Sumerian "Tale of Ziusudra" describes the flood: "The huge boat had been tossed about on the great waters." Contemporary descriptions of the tablet describe a terrible storm that raged for

---

[1] Genesis 7:4
[2] Genesis 7:8
[3] Genesis 7:15–16
[4] https://en. wikipedia. org/wiki/Ziusudra
[5] https://en. wikipedia. org/wiki/Sumerian_King_List#cite_note-18

seven days. It is unclear if the tablets say seven days, or if seven days is merely an incorrect interpretation of Sumerian time units.

"In the six hundredth year of Noah's life, on the seventeenth day of the second month—on that day all the springs of the great deep burst forth, and the floodgates of the heavens were opened. And rain fell on the Earth forty days and forty nights."[1] Notice the word "all" and the plural word "springs" regarding the deep. This suggests that many springs connected the deep to the Earth's surface. In this case, God evidently cut all of them loose.

"Then the flood began. The flood killed all living things that moved on dry land. The waters rose very high… and all the high mountains under the entire heavens were covered. The waters rose and covered the mountains to a depth of more than fifteen cubits."[2]

"But God remembered Noah and all the wild animals and the livestock that were with him in the ark, and he sent a wind over the Earth, and the waters receded."[3] The original Hebrew text uses a word for the English word "wind," meaning "spirit." Therefore, this apparently was not a physical wind but the spirit of God.

"The waters flooded the Earth for a hundred and fifty days."[4] "At the end of the hundred and fifty days the water had gone down, and on the seventeenth day of the seventh month the ark came to rest on the mountains of Ararat. The waters continued to recede until the tenth month, and on the first day of the tenth month the tops of the mountains became visible."[5]

"By the first day of the first month of Noah's six hundredth and first year, the water had dried up from the Earth. Noah then removed the covering from the ark and saw that the surface of the ground was dry. By the twenty-seventh day of the second month the Earth was completely dry."[6]

---

[1] Genesis 7:11–12
[2] Genesis 7:21
[3] Genesis 8:1
[4] Genesis 7:24
[5] Genesis 8:3–5
[6] Genesis 8:13–14

Noah and all its occupants exit the ark. Noah thanks the LORD and offers a sacrifice: "Then Noah built an altar to the Lord and, taking some of all the clean animals and clean birds, he sacrificed burnt offerings on it."[1] The Sumerian "Tale of Ziusudra" recounts the story as well. "Then Utu the sun god appears and Ziusudra opens a window, prostrates himself, and sacrifices an ox and a sheep." After another missing part of the tablet, the text resumes. "After the flood ended, Ziusudra prostrated himself before An[2] and Enlil."[3] In the Sumerian religion, An is known as King of the Gods, Lord of the Constellations, Spirits, and Demons, and Supreme Ruler of the Kingdom of Heaven. Enlil was purported as the only god who could reach An, the god of heaven. The similarities between Mesopotamian records and the Bible are truly amazing.

Flood Story Detractors and Supporters

Scientists like to characterize anything where no physical proof exists as a myth. This is so with the great flood. Nevertheless, a myth is not necessarily false.

Some scientists provide evidence supporting the great flood. In 1851, then noted geologist Edward Hitchcock said, "We ought only to expect that the facts of science, rightly understood, should not contradict the statements of revelation, correctly interpreted… if both records are from God, there can be no real contradiction between them." Hitchcock carried out extensive research supporting the great flood. One observation noted marine fossils transported to high mountain elevations. These fossils seemed to provide clear evidence of submergence under the oceans. Many theories abounded about the date of the great flood, but Hitchcock's work led other scientists to conclude that the great flood occurred five thousand or

---

[1] Genesis 8:20
[2] https://en. wikipedia. org/wiki/Anu
[3] https://en. wikipedia. org/wiki/Enlil

six thousand years ago."[1] Other scientists argue against a great flood, but their work indirectly really does show evidence of a great flood.

Modern scientists have studied changes in the elevation of the Black Sea as a means of finding evidence for the great flood.[2] Modern investigators appear to center their focus on a period much earlier than 4,900 years ago, when the flood happened. Based on data dated approximately eight thousand years ago, these authors do not accept the "Noah's Flood Hypothesis." They cite Black Sea water elevation changed eight thousand years ago, but this does not fit the timeline of the flood narrative. Part of their logic assumes that Bible flood stories place the time of the flood eight thousand years ago. They are completely correct based on this assumption. Changes in elevation of the Black Sea during this period are completely logical, given increasing ocean levels and rainfall due to the melting of the continental ice fields during this time. However, the evidence of these occurrences could not possibly reflect the great flood, especially in magnitude. However, other parts of their work include other clues that support a great flood approximately 4,900 years ago. The Bosporus Strait connects the Mediterranean Sea to the Black Sea. The Bosporus was a freshwater lake until 5,300 years ago. This is during the period when rainfall in the area declined and fresh water flowing into the Black Sea and then on to the Mediterranean declined. This allowed water from the Mediterranean to flow backwards to the Black Sea. Salt water flow into the Black Sea increased astronomically during the great flood, completely covering the Black Sea. Evidence tells science that fresh water started to flow into and out of the Black Sea through the Bosporus approximately 4,400 years ago. As discussed in subsequent sections, this is right around the time waters from the great flood receded back to normal sea levels. Other data and figures in the reference material indicate a sharp increase in the level of the Black sea starting 4,900 years ago. A flood culminated five thousand years ago, resulting in maximum sea levels.

---

[1] R. L. Stiling. *The diminishing deluge: Noah's Flood in nineteenth century American thought.* University of Wisconsin. Madison: Univiersity of Wisconsin, 1991.

[2] Springer. *The Black Sea Flood Question:.* Dordrecht, The Netherlands, The Netherlands: Springer, 2007.

The trade and economic patterns around the Black Sea provides a clear indication of the flood 4,900 years ago. (Springer 2007) Prior to 4,900 years ago very little Black Sea evidence exists showing contact with other civilizations. Immediately after 4,900 years ago, evidence of widespread settlement inland exists. Some settlement appears to support activities around maritime exploitation of resources. Ceramics from Turkey and other regions appears. What accounts for the dramatic change in activity? The Black Sea is near Mount Ararat, where the ark landed. Most likely some of the descendants of people saved in the ark migrated north to the area, instead of south to the old homeland of Mesopotamia.

The spread of languages also provides evidence of the great flood. During the time prior to eight thousand years ago the diversity of language seems considerable.[1] But 4,900 years ago, the diversity was low. The Semitic language is one language that notably gained a strong foothold about five thousand years ago, right after the time of the great flood date. It stands to reason that the Sumerians, in the form of survivors from the ark, continued the language they knew.

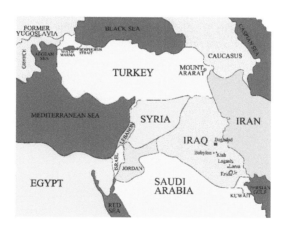

Figure 6-1 – Mediterranean – Black Sea area- Mount Ararat is much closer to the Black Sea than to the old cities of Mesopotamia.

---

[1] Springer. *The Black Sea Flood Question:*. Dordrecht, The Netherlands, The Netherlands: Springer, 2007

Confirming the Science of the Deep

Detractors often raise doubts about the volume of water required to flood the Earth, as stated in the Bible. Scientists seem predisposed to cast doubt on the flood. It is physically impossible to flood the Earth with a four to five-kilometer-deep layer of water.[1] There is not enough water in or on the Earth to reach the tops of mountains. Even with the melting of the polar ice caps, there is far too little water for a boat to land on a mountain. Based on knowledge available over the last few centuries, these criticisms are valid. However, I say, based on recently discovered knowledge, this criticism is not valid.

The Bible mentions the "springs of the deep" in conjunction with the rain that started the flood. The deep is the only source with sufficient water to flood the Earth in the manner described in the Bible. The deep is a recently discovered vast reservoir of water that exists 410 to 650 kilometers below the Earth's surface. I estimated that the great flood needed 4.52 billion cubic kilometers of water, to cover the Earth, as described in the Bible. This is the volume of water needed to cover the Earth, including Mount Everest. As discussed later, I think the deep contains sufficient volumes of water to flood the Earth, as described. In addition to the deep, other heavenly sources of water were in play. "Now the springs of the deep and the floodgates of the heavens had been closed, and the rain had stopped falling from the sky."[2] The Hebrew word for heavens means outer space. In addition to rain caused by the eruption of the deep, water arrived on Earth from outer space. Perhaps this was in the form of comets. Perhaps these arriving comets impacted the Earth and initiated the opening of the Earth to release the vast volumes of water from the deep.

When I first considered the deep as a source of floodwater, one thing crossed my mind. How did all the water get from 410 to 650 kilometers below the Earth's surface to the Earth's surface? The high pressure below the Earth's surface might push up the water. However,

---

[1] R. Chadwick. *First Civilizations*. Oakville, CT, USA: Equinox Publishing Ltd., 2005.
[2] Genesis 8:2

water is incompressible. When a little water escapes the deep, pressure will decline, and then the water is at a lower pressure and temperature. The temperatures and pressures in the deep are so high that water is past the super critical point. Under these conditions, water acts more like a gas than a liquid. Water in the deep will continue to keep flowing, somewhat similar to air that keeps on escaping from a balloon until the balloon is empty.

The water built up over a period of forty days, so the average flow rate was 109 million cubic kilometers per day. I estimated the flow rate of the Euphrates River at 20.9 cubic kilometers per year, or 0.057 cubic kilometers per day. Thus, the average flow rate of water during the buildup of the flood was equivalent to the flow rate of 1,975,000 River Euphrates. This is a staggering flow rate of water.

The Bible states that the flood started "on the seventeenth day of the second month."[1] "On the seventeenth day of the seventh month the ark came to rest on the mountains of Ararat."[2] The time span between the start of the flood and the time the ark hit land is approximately five months or 150 days. This means that it took 110 days for the water to recede from maximum levels to a point where the ark hit ground. The elevation of the upper peak of Mount Ararat[3] is 5,137 meters. About 1.89 billion cubic kilometers of water must return to the deep, to drop water level from above Mount Everest to the top of Mount Ararat. The water recession rate was 17.2 million cubic kilometres per day. This is over six times slower than when the Earth flooded.

"The waters continued to recede until the tenth month, and on the first day of the tenth month, the tops of the mountains became visible."[4] Notice here that the ark came to rest on Mount Ararat, but it took more time for the tops of the mountains to become visible. This means that the ark probably came to rest on the highest peak in the Mount Ararat range; otherwise the tops of

---

[1] Genesis 7:11
[2] Genesis 8:4
[3] https://en. wikipedia. org/wiki/Mount_Ararat
[4] Genesis 8:5

some mountains would become visible before the ark came to rest. The lower peak of Mount Ararat is 3,896 meters in elevation. The volume of water recession required for the lower peak to become visible is 619 million cubic kilometers, which took seventy-five days to accomplish. The water flow rate has dropped to 8.36 million cubic kilometers per day, or over thirteen times slower than when the water level rose.

"By the first day of the first month of Noah's six hundred and first year, the water had dried up from the Earth. Noah then removed the covering from the ark and saw that the surface of the ground was dry."[1] The passage refers to what Noah witnessed from the ark. The general elevation of Turkey in the Mount Ararat area is quite high. Thus, the water had not dried up from the entire Earth, but only around the area where the ark landed. Water over the globe, in general, was still very high and had a long way to drop. Although dry land appeared, the conditions were apparently not ideal. The ark's occupants did not exit the ark until another fifty-six days. That is how long it took the land to become completely dry. During the fifty-six days, residual water on the Earth was likely draining towards the oceans, as the higher mountainous regions of the Earth dried. This is much the same as when we have a heavy rain in modern times. The flow rate in the rivers stays high for days after the rain stops, as the water from the land flows to the rivers and then to the oceans.

"By the twenty-seventh day of the second month, the Earth was completely dry."[2] The initial indication of dry land applies to the land near the ark, but the rest of the Earth was still quite waterlogged and flooded. The flow rates for water recession are very much slower than the average flow rate when the water level was rising. This makes sense from a scientific point of view, as the depth of water on the Earth decreased with time.

Scientists[3] know that Olivine is the base rock comprising the deep and that it holds water. Using this data, I was able to estimate

---

[1] Genesis 8:13
[2] Genesis 8:14
[3] D. G. Pearson. Hydrous mantle transition zone indicated by, 2014.

the volume of water in the deep. As well, there are many other sources of stored water beneath the Earth's surface, starting in the lithosphere just under the oceans and extending into the lower mantle of the Earth.[1] The primary storage region is the transition zone, 410 to 670 kilometers beneath the Earth's surface.

My work leads me to believe that there is sufficient water in the Earth for the great flood, not to mention any water arriving from space. As well, the release of water from the deep likely involves a very explosive process as the rock in the deep responds to changing pressure. As pressure declines in the deep, the rock changes form, resulting in very violent movement, similar to when an earthquake occurs.

I earlier discussed the flow rate of water required to flood the Earth, according to the timeline described in the Bible. In my work as a mechanical engineer, I often do fluid flow calculations to size pipes. I tried a rough calculation, assuming that the fountains of the deep were a series of pipes. This resulted in a huge technical roadblock. The flow rates were so astronomically high that even very large numbers of pipes, miles in diameter, did not yield a solution. Calculations involving flow through porous beds yielded even worse results.

A closer reading of the Bible provided answers. Genesis 7:11 uses two words to describe how the fountains of the deep opened. The Hebrew words for "burst open" has a connotation of to smash open or separate a fissure or to rupture or sever. The word also implies breaking apart into fragments or splitting apart. The first Hebrew word meaning "burst open" adds much meaning to what really happened. A very violent opening of the deep initiated the flow of water through the springs of the deep. This solved the fluid flow problem I encountered. The LORD God apparently carried out the largest frack job on Earth, thousands of years ago.

The Bible mentions the flood for the first time in Genesis 7:4. In this verse, rain is the operative word, describing the cause of the flood. In chapter seven, verse eleven, the Bible also mentions the

---

[1] American Geophysical Union. *Earth's Deep Water Cycle*. Washington, DC, USA: American Geophysical Union, 2006.

"fountains of the deep" as well as the "windows of heaven" as the source of water. Chapter seven, verse twelve again mentions the flood, but in terms of rain only. It seems like rain is the mechanism that flooded the Earth, close to the ark. The "fountains of the deep" were a necessity to provide adequate volumes of water, but God seems to have allowed only rain to fall on the ark. There is a very good reason for this. The waters of the deep are very hot, with temperatures of nearly 1,500 degrees centigrade at the top of the deep and over 1,600 degrees centigrade at the bottom of the deep. Calculations indicate water from the deep emerges partially as steam at the Earth's surface. The steam is very hot and likely heated up the atmosphere to high temperatures, near the point that it emerged. This steam eventually cooled in the atmosphere and condensed into rain that fell back to Earth. It was very important that the fountains of the deep did not erupt anywhere near the ark. Any eruption near the ark likely would scald and kill the ark's inhabitants.

After the Flood: A Long Drying Process

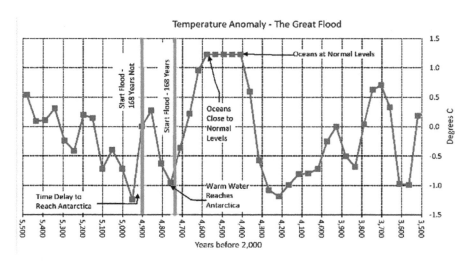

Figure 6-2 – Antarctica ice core data during the time of the flood

After the flood, Noah and his family ended up on the top of a high mountain in present-day Turkey. The Bible ends the story of the flood approximately one year after it started. But it took several hundred years longer for the waters to drop to another 3,900 meters to reach normal sea levels. During this time, much of Mesopotamia was under water. Ice core records help to show this process, 4,900 years ago, when the flood started. Figure 6-2 shows that it took very little time for temperatures in Antarctica to increase. Eruptions from the springs of the deep contained steam and were very hot. This heated up the atmosphere near them. Wind currents on Earth then moved the heat around the globe, causing the atmosphere in Antarctica to heat up relatively quickly. The Antarctic atmosphere stayed heated for two hundred years and then started to drop 4,850 years ago.

Water streams rising from the deep to the Earth's surface heated up everything on the Earth's surface near them, including the oceans. Water returning to the ocean from land heated up the ocean as well. This did not immediately affect the temperature of the ocean around Antarctica. It took a much longer time for ocean currents to move the hot water to Antarctica. About 4,750 years ago, the temperature at Antarctica started to rise again, as shown in figure 6-2. Approximately 4,580 years ago, Antarctic temperatures started to drop again. Subsequent sections will show science tells us that oceans reached normal preflood levels during this time.

I believe that a major return conduit to the deep exists in Antarctica. Although this resulted in huge return flow rates to the deep, complete recession of the floodwaters took hundreds of years. The reason for this lies in the structure of Antarctica.

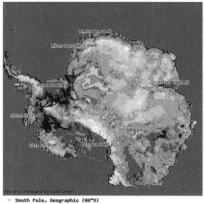
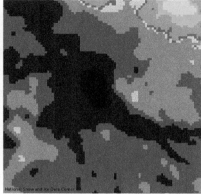

Figure 6-4 Example of a deep hole Near the Ross Ice Shelf in Antarctica

Figure 6-3 – Antarctica Bedrock Profile

Figure 6-3[1] shows a map illustrating the bedrock elevations under the continent of Antarctica. There are some locations with elevations lower than two thousand meters below sea level. One such location is between the Ross Ice Shelf and the Abbot Ice shelf, shown in [2]figure 6-4. I feel that locations such as these are the tops of conduits, returning water to the deep. One such area is approximately eighty kilometers wide and 150 kilometers long. There is no indication of the lowest elevation of this area.

---

[1] http://nsidc. org/data/atlas/news/bedrock_elevation. html- Lythe, M. B., D. G. Vaughan, and the BEDMAP Consortium. 2000. BEDMAP - bed topography of the Antarctic. Cambridge, United Kingdom: British Antarctic Survey. Digital Media. Accessed 28 June 2008. – Public Domain – Use per HSIDC conditions -https://nsidc. org/about/use_copyright. html

[2] Taken from and enlargement of a portion of Figure 6-1

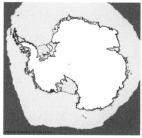 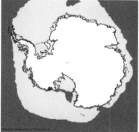 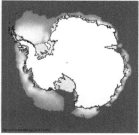

Figure 6-5 Sea ice around Antarctica in June **Figure 6-6 – December sea ice decreases more near suspected eyes of the deep** **Figure 6-7 – Open water appears near suspected eyes of the deep in January**

Figures 6-5 to 6-7 shows sea ice surrounding Antarctica.[1] The extent of sea ice in June reflects the coldest part of the year in Antarctica. In December, temperatures have been rising for some time. It is interesting that most of the melting occurs from the extremities of the sea ice inward, toward land. There are exceptions to this rule, most notably at the Ross Ice Shelf. Here we see that ice melts near land and the ice shelf first. Other locations show indications of early coastal melting, although not as pronounced as at the Ross Ice Shelf. Many of the locations of early coastal melting are near areas where bedrock elevation drops down significantly and some are near ice shelves. Why does sea ice near the ice shelves melt first? Possible geology of the bedrock under the Ross Ice Shelf may explain the phenomena.

---

[1] Image/photo courtesy of the National Snow and Ice Data Center, University of Colorado, Boulder, link lost

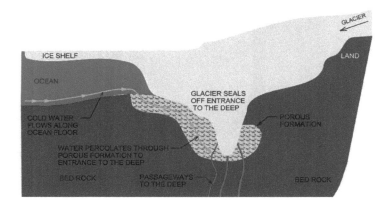

Figure 6-8 - Water moving into the passageways to the deep under an ice shelf

The sea bed in western Antarctica and around the Ross Ice Shelf consists of sedimentary layers.[1][2][3][4] The geologic characteristics at the Ross Seas allow water to percolate toward the passageways to the deep. This explains the early melting of sea ice near the shelf.[5] Very salty and cold Antarctica bottom water surrounds Antarctica. This water very likely flows through the porous formations to reach the passageways to the deep so it can return to the deep. This pulls in warmer surface water under the ice shelf, causing melting closer to land.

During the flood, water level rose around Antarctica, raising the ice covering these openings to the deep. Now fully exposed, the flow rates to the deep increased drastically. This eventually caused a rapid temperature rise at Antarctica 4,750 years ago, as indicated in figure 6-2. As ocean levels dropped, ice dropped, eventually covering the

---

[1] Geopress. *Techtonic, Climatic and Cyrospheric Evolution of the Antartci Peninsula.* (J. B. Welner, Ed.) Washington, DC, USA: American Geophysical Union, 2011.
[2] Oxford Sciences Publications. *The Geology of Antarctica.* (R. J. Tingey, Ed.) Oxford, NY, Usa: Oxford University Press, 1991.
[3] https://en. wikipedia. org/wiki/Ross_Sea
[4] https://en. wikipedia. org/wiki/En_echelon_veins
[5] https://en. wikipedia. org/wiki/Antarctic_Bottom_Water

openings to the deep. When fully covered, flow to the return conduits, again flowed through the porous layers, at a greatly reduced flow rate.

Answers on the 168 Lost Years

Figure 6-2 sheds some light on the supposed 168 lost years in the Hebrew calendar. There is speculation that the Hebrew calendar lost 168 years during the captivity in Persia. The first red line in figure 6-2 is the calculated start of the flood, without 168 lost years. The first line coincides with the geological ice core markers, showing pronounced temperature increase and start of the flood. The first line correctly matches the start of the flood. The line including 168 missing years completely misses the correct date of the flood. Thus, the Hebrew calendar, likely, did not lose 168 years, based on these criteria.

Noah's Sons and Grandchildren: The First Kings after the Flood

"The sons of Noah who came out of the ark were Shem, Ham and Japheth."[1] The next verse was quite prophetic regarding the relationship between the descendants of Noah's sons. "These were the three sons of Noah, and from them came the people who were scattered over the whole Earth."[2] The Hebrew word used in the original Bible text for the word "scattered" means "shattered into pieces," and to be "widely dispersed." The wording of this verse hardly projects an amicable parting of the ways. History and world events, even today, have borne this out.

The ark landed on Mount Ararat, in Turkey. The distance between the ark's starting point and Mount Ararat is over a thousand kilometers. Noah's descendants eventually made it back to Mesopotamia, but it took a long time. Not only was the distance great and treacherous but the waters of the flood took a long time to recede.

---

[1] Bible Genesis Chapter 9 Verse 18
[2] Bible Genesis Chapter 9 Verse 19

The Sumerian King List shows kings after the flood. Since all persons except Noah and his family perished in the flood, these post flood kings had to be from Noah's line. The king list tells us that the post flood kings did not take up reign in Mesopotamia until a few hundred years after the flood. As ocean levels dropped, previously flooded ancient cities reappeared. Post flood kings located their kingships in the old Mesopotamian cities, as they reappeared. The Sumerian King List again states that kingship descended from heaven after the flood. Jushur was the first king of Kish.[1]

The Sumerian King List defines King Enmebaragesi as the twenty-first king to reign after the flood. Two fragments of alabaster vessels found at Khafajah in the Diyala region of eastern Iraq, one of which calls him lugal "king of Kish," bear his name.[2] These fragments independently confirm that King Enmebaragesi's rule started approximately 4,600 years ago.[3] The king list identifies all 21 kings, including the length of time they ruled, between Enmebaragesi and Jushur. When I apply my derived factor to convert sars to Gregorian time, I calculate that Jushur's reign started 4,750 years ago, nearly 150 years after the flood. Only an event of monumental proportions such as the great flood would interrupt the kingship of the area for 150 years. Combined with the scientific evidence in figure 6-2, the evidence appears overwhelming.

The location of Kish[4] is in the current day Babil Governorate in Iraq. The capital of Babil Governorate is present day Hillah.[5] Hillah has an elevation of thirty-four meters above sea level. Inhabitants reoccupied Kish, 150 years after the flood. Thus, it took 150 years for the flood waters to drop to thirty-four meters above sea level.

Some scientists believe that the second listed king is really an inscription that means, "All of them (were) lord," and there really

---

[1] S. Langdon. *Oxford Editions of Cuneiform Texts*. OXFORD UNIVERSITY PRESS, 1923.
[2] D. Potts. *The Arceology of Elam*. Cambridge, UK: Cambridge University Press, 2004.
[3] https://en. wikipedia. org/wiki/Sumerian_King_List
[4] https://en. wikipedia. org/wiki/Kish_(Sumer)
[5] https://en. wikipedia. org/wiki/Hillah

was no formal kingship for a time. This makes much of sense. The world's population started out at six, after the flood. In 150 years, the population probably increased to a few thousand at most. There was no need for an official kingship and kingdom. The numbers of people did not warrant a formal societal structure. As well, everyone spoke the same language. Karen Rhea Nemet-Nejat[1] shows how governing structures changed before and after the flood. Before the flood, the temple was the center of social, economic, and political life. The temple controlled the life of residents. After the flood city states were loosely organized, exercising little formal control. Undeveloped dependent populations surrounded each urban city-state.

The Temple of E-ana at Uruk: The Tower of Babel

The Sumerian King List displays King Meshkianggasher as the first post flood king of the City State of Uruk, after Jashur ruled in Kish. Meshkianggasher became king of Uruk 4,600 years ago, 150 years after the rule of Jashur started. Archaeologists tell us ancient stone tablets have Meshkianggasher,[2] as the first ruler of Uruk, associated with Eana. The "e" in Eana[3] is the Sumerian word or symbol for house or temple. Eana, known as the "house of heaven" was the name of the temple dedicated to Inanna at Uruk. The tablets tell us that Meshkianggasher's son was to build the city of Uruk, around the temple of Eana. He intended the temple as the main temple to its patron goddess Inanna.[4]

Chapter 11 of Genesis tells us that "as people moved eastward, they found a plain in Shinar and settled there."[5] Shinar[6] [7] is an area in Mesopotamia that encompasses what would eventually become

---

[1] K. R. Nemet-Nejat. *Daily Life in Ancient Mesopotamia*. Westport, CT, USA: Greenwod Press, 1998.
[2] https://en.wikipedia.org/wiki/Mesh-ki-ang-gasher
[3] https://en.wikipedia.org/wiki/%C3%89_(temple)
[4] https://en.wikipedia.org/wiki/Inanna
[5] Genesis 11:2
[6] https://en.wikipedia.org/wiki/Shinar
[7] http://jewishencyclopedia.com/articles/13582-shinar

Babylonia. The area encompasses both Babel in northern Babylonia and Erech (Uruk) in southern Babylonia. The people try to build the Tower of Babel. The LORD decides to confound their efforts by confusing their language. The people dispersed from there, each group with a different language.

Could the great Temple of Eana be the Tower of Babel? I think the answer is yes. The location of Eana is in the same area as the proverbial Tower of Babel. As well, stories on ancient stone tablets tell of events that mirror tower of Babel events in the Bible. "Enmerkar and the Lord of Aratta"[1] is a Sumerian story telling of conquests between Enmerkar, king of Uruk and the Lord of Aratta. The story includes an account of the Sumerian "confusion of tongues."[2] It also involves Enmerkar constructing temples at Eridu and Uruk. Scholars consider the confusion of tongues as a myth that resulted in the Tower of Babel verses in the Bible. As I indicated before, myths are stories not proven by scientific fact; but myths are often true.

The Recurring Competition among Men

Contemporary commentaries on the Sumerian King List mention no conquests until the reign of the second last king of the First Dynasty of Kish, named Enmebaragesi. Enmebaragesi's[3] rule started approximately 4,600 years ago. Enmebaragesi captured Elam,[4] an area in present-day Iran. Again, man's nature drove him to fight with his neighbours.

The king list has the following inscription for city-state of Kish: "Then Kish was defeated and the kingship was taken to E-ana." The post flood Kingdom of Uruk or Ur did not last long. After about fifteen years, the Kingdom of Ur moved to Awan[5]. The Awan Dynasty lasted thirty years before returning to Kish. The second dynasty of

---

[1] https://en. wikipedia. org/wiki/Enmerkar_and_the_Lord_of_Aratta
[2] https://en. wikipedia. org/wiki/Confusion_of_tongues
[3] https://en. wikipedia. org/wiki/Enmebaragesi
[4] https://en. wikipedia. org/wiki/Elam
[5] https://en. wikipedia. org/wiki/Awan_dynasty

Kish lasted 175 years. All three of these city-states fell by about 4,400 years ago.

The elevation of the city of Ur[1] is four meters above sea level. Thus, between 4,700 and 4,600 years ago, the ocean level dropped five meters. Figure 6-2 shows that Antarctic temperature stabilized during this period. Sea level dropped to within four meters of normal. Antarctic ice nearly covered the eyes of the deep, restricting flow rate back to the deep.

The tablets of the Sumerian King List continue, listing many kings. The king list ends with the inscription "Then Urim was defeated. The very foundation of Sumer was torn out. The kingship was taken to Isin." This apparently marks the end of the Samarians, as a ruling class. However, the LORD had other plans to carry on his line.

Abraham: The New Root of the LORD's People

Man's story has been and still is one of continual disobedience to the LORD God, resulting correction and punishment. The creation of Adam, Adam's dominion over the Earth, the flood, and the eventual end of the Semitic culture are examples of this. However, the LORD always finds a righteous remnant of his people to start fresh with.

The LORD promised to never again to send a flood, to wipe out the world. His promise reveals a telling aspect of man's nature: "Never again will I curse the ground because of humans, even though every inclination of the human heart is evil from childhood."[2] The LORD tells us that evil inclines man's thoughts right from birth. And so it is that man's nature continuously causes him to go astray.

The Semite line carried on through the descendants of Shem, one of Noah's sons. Eventually the line descended to Abram, otherwise known as Abraham. The birth of Abram occurred around 4,600 years ago. The arrival of Abram was during the end of the First

---

[1] https://en. wikipedia. org/wiki/Ur
[2] Genesis 8:21

Dynasty of Kish. Abram is the tenth generation of descendants from Noah, since the flood.

Abraham was the remnant chosen by the LORD to carry on his line on Earth. God again chose a single worthy person and his family to procreate a new line of human beings to further his plan on Earth. The LORD promised Abram that his descendants would become a great nation:

> The LORD said to Abram, "Go from your country, your people and your father's household to the land I will show you. I will make you into a great nation, and I will bless you; I will make your name great, and you will be a blessing. I will bless those who bless you, and whoever curses you I will curse; and all peoples on Earth will be blessed through you."[1]

From this point onward, the Bible shows Abraham travelling from one location to another. The line of Abraham eventually resulted in Moses. Moses was a key figure in the LORD's plan. The LORD God called Moses up to Mount Sinai. The LORD God talked directly to Moses and gave him the Ten Commandments. Even before Moses could walk down from the mountain, his people started worshipping false gods. Therefore, the cycle went on. The LORD needed to discipline his people again. He put them into exile.

Nevertheless, the LORD brought the Hebrews back from captivity. His descendants again eventually took over the promised land and became the remaining tribes of Judah. We see that modern Israel is alive today. Modern Israel descended from Abraham.

---

[1] Genesis 12:1-3

Semitic Laws, Institutions, and Traditions: Lasting Legacies

The Sumerians left a legacy that endures even to present times. Many Bible stories and modern laws have similarities ancient Sumerian stories and codes.[1]

Mesopotamians suffered under taxation, just like we do today. President Trump's election in the USA preceded the time before I wrote this section. One of President Trump's election promises included reduction of taxes. As such President Trump is a modern day Urukagina. Urukagina served as the king of the city state of Lagash, about 4,500 years ago. Urukagina reduced the burden of government on the people. The Code of Urukagina limited the power of the priesthood and large property owners. It took measures against usury, burdensome controls, hunger, theft, murder, seizure of people's property, and slavery. Urukagina stated, "The widow and the orphan were no longer at the mercy of the powerful man." But the more things change, the more they stay the same. Urukagina removed corrupt civil servants and cleared the city of criminals. But as with today's politicians, reforms soon disapeared after Urukagina's reign ended in less than ten years.[2]

The Code of Ur-Nammu[3] is one of the first legal codes in history. Science credits the king of Ur with enacting the code some four thousand years ago. Thus, even though the line of Abraham apparently departed from the early dynastic kings, these kings still possibly affected laws in the Bible years later. The Code of Ur-Nammu contains laws regarding murder, robbery, perjury, theft, fraud and other matters. The code even had similarities to the biblical "eye for an eye and tooth for a tooth." According to Kramer, this law was circumvented in Mesopotamia with more humane money fines instead.

According to Kramer, political assemblies also had their start in ancient Mesopotamia. In 2800 B.C. a government with a "house"

---

[1] S. N. Kramer. *History Begins at Sumer.* USA: Falcons Wing Press, 1959.
[2] S. N. Kramer. *History Begins at Sumer.* USA: Falcons Wing Press, 1959.
[3] https://en. wikipedia. org/wiki/Code_of_Ur-Nammu

and a "senate" evolved. The house and senate argued between policies of war and arms-bearing and appeasement involving peace. The king sometimes played politics and promoted one body over the other to suit his own views.

Schools are a Mesopotamian institution. The Sumerian school originally educated scribes required to satisfy the economic and administrative demands of the land, primarily those of the temple and palace. However, schooling eventually became increasingly wide spread.

Sumerians produced technical manuals as well, such as the equivalent of a farmer's almanac and pharmaceutical manuals. As well, Mesopotamian writings reflect various texts of the Bible such as, the book of Job, Proverbs, the Flood of Noah, and the resurrection.

7

Living Things and Evolution

This chapter looks at the evolution of living things on Earth. It provides a summary of the progression of life, starting from single-celled organisms billions of years ago to present-day man. The discussion focuses on how mammals evolved. Science tells us that humans evolved from mammals, more specifically primates or monkeys. The Bible covers in detail only the last fifteen thousand to twenty thousand years or so of the history of the world. As discussed, many mass extinctions of life occurred prior to the time of Genesis.

Many creationists believe that God created every living thing from nothing. Nothing could be more wrong. Nature used the DNA of the first simple multicelled organisms as a starting point for all living things on Earth. When the LORD formed certain types of animals in chapter 2 of Genesis, he really created nothing. He used the dust of the Earth or possibly other existing animals as a building block or starting point for new life. The change required only a very slight modification of the DNA in existing life. Farmers use somewhat the same process when selectively breeding animals with advanced traits.

The Biblical Jump in Time

Modern translations of the first verse in Genesis start by referring to the creation of the heavens and Earth. As we showed in previous chapters, the original Hebrew Bible tells us that no creation happened at all. Only a selection process occurred. The time was only about thirteen thousand years ago. Thus, the first verse of Genesis skips about four billion years of the Earth's development, including the beginning of life forms.

We will see in chapter 8 that the Bible briefly talks about the very beginning of the universe, the Big Bang. However, for the most part, the Bible skips most of the history of the Earth and universe, between the Big Bang and thirteen thousand years ago.

Science tells us that the Earth endured many catastrophic events after its initial formation during the Big Bang. Yet the Bible mentions only one such catastrophic event, the flood in Noah's day. The second verse in chapter 1 of Genesis indirectly refers to a previous extinction event when it tells us that the Earth became formless and empty, before the start of the six days of God's work. Why does the Bible not mention previous catastrophic events? Why is the flood of Noah the only event described in detail, in the Bible? The reason is that biblical writers did not intend to cover previous events on Earth. They intended the Bible to cover only the last fifteen thousand to twenty thousand years or so of history.

Scientists have documented only about 1.2 million of the ten million to fourteen million species that evolution[1] produced. The Earth has not been kind to all species. Of the all the species that ever existed, 99 percent are extinct. The Earth has undergone many events since its formation[2] that drastically changed its surface and affected life.[3] Most of living species faced extinction by Mother Nature. While species become extinct, their DNA does not (Department of Ecology and Evolutionary Biology - University of California n.d.). Extraction of prehistoric DNA from fossils is a science today. Scientists can extract DNA from fossils[4] and so nature probably can too. I know of no science that examines if nature uses DNA from extinct species to reestablish new kinds of life after an extinction. This is an interesting question, which I hope science starts to examine.

---

[1] https://en. wikipedia. org/wiki/Evolutionary_history_of_life
[2] https://en. wikipedia. org/wiki/Extinction_event
[3] V. Coutillot. *Evolutionary Catastrophes: The Science of Mass Extinction.* Cambridge, UK: Cambridge University Press, 2003.
[4] https://en. wikipedia. org/wiki/Ancient_DNA

Scientists disagree about an event called "Snowball Earth." Some believe that about 650 million years ago, the Earth cooled off. Some scientists believe that ice or snow covered most of the globe during this period, even close to the equator.

About 252 million years ago, the Permian-Triassic extinction event occurred, causing a 90 percent extinction of all marine, vertebrate, and insect species. The Cretaceous-Paleogene extinction event some sixty-five million years ago wiped out 80 percent of all species on Earth, including the dinosaurs.[1]

The Quaternary extinction event is one of the last major natural extinction events. Scientists think that the Quaternary extinction event started about fifty thousand years ago in Australia, sixty thousand years ago in North America, and eighty thousand years ago in Northern Eurasia.[2] The Quaternary Extinction Event was especially severe in Australia, where fifteen out of sixteen types of mammals became extinct. Next hardest hit were the Americas, which lost thirty-three out of forty-five types of mammals in North America and forty-six out of fifty-eight types of mammals in South America. Europe and Sub-Saharan Africa fared better, with losses of seven out of twenty-three and two out of forty-four respectively. Scientists attribute the Quaternary Extinction Event to climate change, over hunting by humans and changes in habitat. There is some disagreement amongst scientists regarding the over hunting theory.

Science sometimes classifies the Holocene extinction as part of the Quaternary Extinction Event.[3] The Holocene extinction started between nine thousand and thirteen thousand years ago. During this time, many large mammals disappeared. Scientists consider the Holocene extinction as still in effect. Proponents attribute the causes mainly to man. It is interesting that the great flood during Noah's time happened shortly after this period started. In geologic time scales, the great flood is very close to this period. The Bible mentions none of these events, except the great flood.

---

[1] V. Coutillot. *Evolutionary Catastrophes: The Science of Mass Extinction.* Cambridge, UK: Cambridge University Press, 2003.
[2] http://www. wikiwand. com/en/Quaternary_extinction_event
[3] http://www. wikiwand. com/en/Holocene_extinction

DNA and RNA and Evolution

All complex living plants and animals have DNA to govern their makeup and living processes. DNA has developed over billions of years, ending with the blue print for the modern plants, animals, and humankind.

All living plants and animals have a genetic blueprint that determines how they grow and live. We find this blueprint in the organism's DNA. This applies to all complex living things, except possibly single-celled organisms. RNA is a molecule that reads DNA, causing the organism to reproduce, grow, and live according to its DNA blueprint. Some scientists believe that the first simple organisms skipped the DNA blue print and that RNA only was responsible for replicating the organism.

When man sets out to construct a large building, architects analyze what they need to build. Calculations and analysis determine how to construct the building safely and in a functional manner. A set of drawings and specifications then result from their efforts. This allows suppliers, construction crews, and others to construct the building correctly. In the case of future additions or modifications to the building, existing drawings and specifications form a starting point and architects and engineers revise them as required. For new similar buildings, existing drawings and specifications serve as a starting point to make new drawings and specifications. Engineers and architects always fall back on existing knowledge, to construct new things. They generally do not start from scratch, if previous work is applicable to new ideas.

It makes sense that the same process takes place for the formation of living organisms on Earth. Just as engineers and architects take time for a design phase before construction, nature needed time to develop DNA blueprint molecules and RNA builder molecules. In the same way, nature does not start from scratch to evolve new organisms. Nature starts with existing DNA and RNA and builds on it. This is obvious when examining DNA and RNA. DNA and RNA for all living things on Earth have similarities, which indicate evolution from a common source.

Nature's evolution of DNA and RNA blueprints is more efficient than man's drawings and specifications for buildings. Nature's first design for single-celled organisms was so well thought out that it served as the starting point for all living organisms on Earth. Man must start from scratch when designing buildings and other things significantly different from those already in existence. I am glad that man is not as advanced as nature. As an engineer, I would be out of a job if the first set of drawings and specifications ever developed formed the starting point for all things on Earth.

Science espouses that evolution, on its own, advanced living organisms on Earth. I do not espouse that this theory is invalid. However, think of it for a while. If this theory is correct, it glorifies the power and infinite intelligence of the LORD. When engineers design a building, unexpected conditions sometimes arise, requiring adjustments to the original plan. If evolution by itself resulted in life as we see it today, then the original blueprint for DNA for early organisms was the greatest master plan imaginable. Alternately, maybe the LORD God stepped in along the way to nudge evolution in the right direction.

When did nature develop the DNA and RNA blueprint for living things on Earth? Obviously before the living things on Earth appeared. There must be a period before the formation of the first single-celled organisms, where nature developed the DNA blueprint. The first blueprint may have included RNA only, if scientists are correct that the first single-celled organisms were self-replicating and required RNA only. Otherwise, nature had to develop both DNA and RNA in the original design process.

The question of how DNA formed from the rocky beginnings of the Earth is a question that perplexes scientists. Like the Big Bang, there is no way to replicate the process of DNA molecule formation. One of the building blocks of DNA is amino acids. Scientists cannot create amino acids from inorganic compounds, such as those that formed the Earth billions of years ago. So where did these building blocks and DNA come from?

Spacecraft with mass spectrometers have found these compounds in some comets. Although findings are preliminary, scientists have found more than eighty elements or compounds. These include simple hydrogen to complex amino acids, the building blocks of proteins and DNA. As well, some contain a variety of other compounds. Interstellar dust left over from the Big Bang contains all but one of the compounds found in comets: sulfur.[1] It appears the LORD created these compounds during or after the Big Bang.

The First Life on Earth

Figure 7-1 Modern-day microbial mats in Shark Bay, Australia

Life started on Earth with simple single-celled, bacteria-like organisms. Upon initial formation, the Earth was a rocky place with no dust or soil. Scientists think the first simple organisms on Earth were responsible for breaking the rock surface into dust and soil. For more than two billion years, these simple organisms laid the way for other more advanced plant species that required soil.[2] (S. R. Palumbi 2014) Scientists discovered the earliest microbial mats in Australia and Africa in 3.4-billion-year old rock.[3] Photosynthesis

---

[1] D. J. Eicher. *Comets - Visitors From Deeppace*. New York: Cambridge University Press, 2013.
[2] V. Coutillot. *Evolutionary Catastrophes: The Science of Mass Extinction*. Cambridge, UK: Cambridge University Press, 2003.
[3] S. R. Palumbi. *The Extreme Life of the Sea*. Princeton, NJ, USA: Princeton University Press, 2014.

around 3.5 billion years ago eventually led to a buildup of a waste product, oxygen, in the atmosphere. This led to the great oxygenation event, beginning around 2.4 billion years ago. The emergence of oxygenic photosynthesis in bacteria increased biological productivity by a huge factor.

Simple single-celled organisms lived in layers inside microbial mats. Simple single cells are prokaryotic, which means cells without a membrane-bound nucleus. Organisms with prokaryotic cells reproduce asexually, meaning that offspring arise from a single organism. Each child organism inherits the genes of one single parent only. Each layer of the microbial mat provides food for the next layer. Figure 7-1[1] shows stromatolites in Shark Bay, Western Australia. Stromatolites are an example of modern-day microbial mats. The top layer of stromatolites consists of photosynthesizing cyanobacteria, which create an oxygen-rich environment.

Vaalbara, the beginning of the first known super continent, was in place at the time microbial mats started on the Earth. Vaalbara initially consisted of many small landmasses in the ocean. Microbial mats seem to thrive best along the shores of oceans. The extensive shoreline makeup of Vaalbara was perfect to maximize the growth of microbial mats. Bombarding asteroids possibly caused the smattering of smaller landmasses and seems to have provided an optimum environment, to maximize the growth of single-celled organisms. The formation of Vaalbara, especially if by asteroid strikes, might seem like a random occurrence, but the way it fits into the progression of life on Earth makes me think that there was some method to the madness.

---

[1] https://en. wikipedia. org/wiki/Evolutionary_history_of_life- Stromatolites by Paul Harrison is licensed underCreative CommonsAttribution-Share Alike 3. 0 Unported

More Complex Organisms as the Earth Changes

Figure 7-3[1] shows more complex multicellular organisms developed seven hundred million years ago.[2]

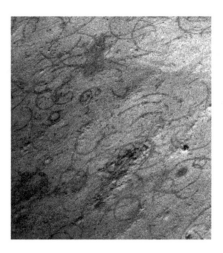

Figure 7-2 Fossil remains of Grypania fungi - one of the first multicelled, oxygen-using organisms on Earth

These organisms have eukaryotic cells, meaning the cells are membrane-bound with a nucleus that contains genetic material. Like single-celled prokaryotic organisms, some eukaryotic-celled organisms reproduce asexually. Most organisms with eukaryotic cells reproduce sexually, meaning that DNA comes from two different parent organisms. Thus, each descendant combines the differing characteristics of its parents. The child becomes something other than an exact copy of its parents. Sexual reproduction allows the process of evolving species to speed up. The number of permutations and combinations of possible

---

[1] https://en. wikipedia. org/wiki/File:Grypania_spiralis. JPG#filelinks- Grypania spiralis by Xvazquez is licensed under Creative CommonsAttribution-Share Alike 3. 0 Unported

[2] V. Coutillot. *Evolutionary Catastrophes: The Science of Mass Extinction.* Cambridge, UK: Cambridge University Press, 2003.

species now increases drastically. As well, some descendants develop bad traits. The process of natural selection weeds out weak descendants as strong descendants thrive. Later Eukaryotes consumed oxygen. Increased biological functioning due to oxygen, along with natural selection, allowed living things to increase at a much faster rate.

Figure 7-3 Francevillian biota fossil

More complex organisms developed. Early eukaryotes were still very basic organisms, but scientists believe that these organisms were the first to use oxygen in their metabolism. The first known eukaryotic organisms were Grypania fungi. Figure 7-2 shows fossil evidence of the Grypania spirals[1]. (Armstrong and D 2005) Grypania spirals appeared 2.1 billion years ago. Other early eukaryotic organisms include the Francevillian biota (fossil remains shown in figure 7-3).[2] Multicellular organisms of very many kinds evolved. The stovepipe sponge in figure 7-4 is one modern example of them[3].

Figure 7-4[3] Modern-day stove sponge – Its prehistoric ancestors were some of the early multicellular organisms

---

[1] https://en. wikipedia. org/wiki/Grypania
[2] https://en. wikipedia. org/wiki/Francevillian_biota
[3] https://en. wikipedia. org/wiki/Sponge

Figures 7-3[1] and 7-4[2] show examples of differentiated multicellular construction that, among other things, allow grouping of similar cells. Similar groups of cells can then assume different functions in the organism. Today our bodies have many groups of similar cells that form the organs and other parts of our bodies. Each of these has a separate job in the functioning of the human body.

Early Animals

Evolution of animal species increased rapidly about 542 million years ago, during a time known as the Cambrian Explosion. This period lasted for about twenty-five million years. Most of the animals that developed during this period were very strange creatures, by today's standards. They were not of great size and often were only a few centimeters long. Some fossilized specimens showed signs of defensive adaptations, such as molluscs, which developed shells. During the end of this period, creatures resembling modern fish evolved, having vertebrae, jaws and fins. During this period most animals and plants lived in water, although some microorganisms may have made it onto land during this time.

---

[1] https://en. wikipedia. org/wiki/Francevillian_biota- oldest multizelluar eucaryotic life by Ventus55is licensed under Creative CommonsAttribution-Share Alike 3. 0 Unported

[2] https://en. wikipedia. org/wiki/Sponge- Aplysina archeri by Nick Hobgood is(Nhobgood (talk)) islicensed under Creative CommonsAttribution-Share Alike 3. 0 Unported

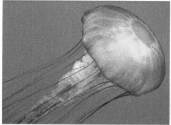  

Figure 7-5 Pacific Nettle Jelly Fish – Modern descendent of ancient jelly fish

Figure 7-6 Modern day arthropod

Figure 7-7² Fossil of ancient arthropod

Some of the earliest animals are jellyfish. Many species of jellyfish appeared about 580 million years ago. Figure 7-5[1] shows a modern jellyfish descendent, the Pacific Nettle Jellyfish. Figures 7-6 and 7-7[2] show modern and extinct arthropods.

Some theories state that species in the animal kingdom increased rapidly in the Precambrian Explosion. Early animals during this period included the Opabinia, in figure 7-8.[3] These creatures lived in the sea and had no vertebrae. (S. R. Palumbi 2014) Opabinia likely trolled the depths of the muddy bottoms of the Cambrian Sea, hunting for small creatures. It

Figure 7-8 Opabinia – Ancient sea animals without vertebrae

---

[1] https://en. wikipedia. org/wiki/Jellyfish#Taxonomy- A Pacific sea nettle by Dan90266 and https://www. flickr. com/photos/dan90266/37269957/is licensed under Creative CommonsAttribution-Share Alike 2. 0 Generic

[2] https://en. wikipedia. org/wiki/ArthropodA Arthropoda collagebyGuy Haimovitch andPeter Halaszis licensed under Creative CommonsAttribution-Share Alike 3. 0 Unported- File cropped from original multiple image

[3] https://en. wikipedia. org/wiki/File:Opabinia_BW2. jpg - Opabinia regalisbyNobu Tamura (http://spinops. blogspot. comis licensed under Creative CommonsAttribution-Share Alike 3. 0 Unported

was only two inches long. All of today's modern creatures have ancestors from the Cambrian Explosion. The oceans sheltered life. Much later life advanced and moved to land.[1] The Precambrian Explosion is a very important part of the evolutionary chain.

First Vertebrae Animals

Approximately 600 million years ago, organisms developed skeletons.[2] Figure 7-9[3] shows a depiction of a Haikouichthys, which arrived approximately 525 million years ago. The Haikouichthys is one of a number of early fish like animals with vertebrae. These animals did not have jaws, and filter fed themselves, near the seabed. The Haikouichthys was about 2.5 cm long.

Figure 7-9 Haikouichthys – One of the first animals with vertebrae

Approximately 450 million years ago, animals appeared such as the Acanthodians, shown in figure 7-10.[4] These are among the first known animals with a backbone and jaws. Proponents often call them tiny sharks.

Figure 7-10[3] Acanthodians – One of the first animals with a backbone and jaws

---

[1] Palumbi, S. R. (2014). *The Extreme Life of the Sea*. Princeton, NJ, USA: Princeton University Press.

[2] Coutillot, V. (2003). *Evolutionary Catastrophes: The Science of Mass Extinction*. Cambridge, UK: Cambridge University Press.

[3] https://en. wikipedia. org/wiki/Vertebrate - The primitive vertebrateHaikouichthys by Nobu Tamurais licensed under Creative CommonsAttribution-Share Alike 3. 0 Unported

[4] https://en. wikipedia. org/wiki/File:Acanthodes_BW. jpg - Acanthodes bronni by Nobu Tamurais licensed under Creative CommonsAttribution-Share Alike 3. 0 Unported

Transition to Land

Figure 7-11 Acanthostega - one of the first animals to move to land

Limbs evolved mostly as a means of locomotion of tetrapods on land. The limb with digits was initially considered a terrestrial adaptation. Early animals such as the Acanthostega developed limbs before moving to land.[1] Arthropods mainly breathe using gills, usually located above their legs. Most land going arthropods evolved to have lungs. Arthropods may have ventured onto land as early as 440 million years ago.

The estimated time of arrival of the earliest amphibious animals is 300 to 330 million years ago. The Acanthostega, shown in figure 7-11[2], is an example of one of the first animals to appear on land. Scientists believe that the Acanthostega evolved from a fish and was poorly adapted for land. Its body was not strong enough to withstand the forces of gravity. It probably lived in shallow water puddles to allow water to partially support its body weight.

The skeleton of *Acanthostega* remained largely as cartilage, a common trait for aquatic tetrapods. This suggests a largely aquatic lifestyle.[3] Around 400 million years ago water dwelling animals developed stronger eggs.[4]

---

[1] M. Laurin. *How Vertebrates Left The Water.* Berkeley and Los Angeles, California, USA: University of California Press, 2010.
[2] https://en. wikipedia. org/wiki/Acanthostega- Acanthostega gunnari by Nobu Tamura(http://spinops. blogspot. com) is licensed under Creative CommonsAttribution-Share Alike 3. 0 Unported
[3] M. Laurin. *How Vertebrates Left The Water.* Berkeley and Los Angeles, California, USA: University of California Press, 2010.
[4] V. Coutillot. *Evolutionary Catastrophes: The Science of Mass Extinction.* Cambridge, UK: Cambridge University Press, 2003.

Amniotes lay eggs equipped with two shell layers, allowing them to lay eggs on land. This freed amniotes from the need to return to water to procreate. The first truly land-living vertebrates appeared during the period during which the oldest amniotes lived.[1] (Kemp 2005) The Westothiana, shown in figure 7-12[2] is one of the earliest amniotes that lived 360 million years ago.[3] Amniotes include reptiles and mammals.[4]

Figure 7-12 – The Westothiana – One of the earliest Amniotes

This is very interesting from a biblical point of view. When expanding the role of domestic animals, God included mammals and reptiles in the groups of animals he chose to expand. A natural link also appears evident in the evolution of these two groups.

When they first appeared, amniotes consisted of the sauropsid group (reptiles) and synapsids (mammals). Synapsids dominated the scene until 250 million years ago, until the great Permian crisis eliminated most of them. Sauropsids, such as birds and dinosaurs, then dominated. The Cretaceous/Paleogene crisis sixty-five million years ago eliminated many large reptiles, including most dinosaurs. Since then, synapsids, represented exclusively by mammals, again dominated the scene.[5]

---

[1] M. Laurin. *How Vertebrates Left The Water.* Berkeley and Los Angeles, California, USA: University of California Press, 2010.

[2] https://en. wikipedia. org/wiki/Westlothiana- Westlothiana lizziae by Nobu Tamura(http://spinops. blogspot. com) is licensed under Creative CommonsAttribution-Share Alike 3. 0 Unported

[3] Kemp, T. (2005). *The Origin and Evolution of Mammals.* New York, NY, USA: Oxford University Press.

[4] M. Laurin. *How Vertebrates Left The Water.* Berkeley and Los Angeles, California, USA: University of California Press, 2010.

[5] M. Laurin. *How Vertebrates Left The Water.* Berkeley and Los Angeles, California, USA: University of California Press, 2010.

The archaeothyris, shown in figure 7-13,[1] evolved about 330 million years ago. It belongs to the general animal classification called "Synapsids," which evolved into mammals. The aarchaeothyris belongs to a subgroup of synapsids named amniotes. Warm blooded animals evolved approximately 260 million years ago.[2]

Figure 7-13 Archaeothyris – one of the first synapsids – the group mammals emerged from

Synapsids evolved into a more advanced group named therapsids. Therapsids developed separate mouth and airway passages, allowing them to eat and breathe at the same time. As well, scientists believe that this feature allowed chewing of food, which speeded up digestion and metabolism. This adaption allowed increased metabolic rates. Another development, the bone that holds the lower teeth, slowly became the lower jaw. Another development was vertical erect limbs versus the more sideways sprawling limbs of their ancestors. This increased the mobility of therapsids.

Figure 7-14 Inostrancevia- One of the first Therapsids – A group with an upright, rather than a sprawling stance

---

[1] https://en.wikipedia.org/wiki/Archaeothyris-*Archaeothyris*by User:ArthurWeasleyis licensed under Creative CommonsAttribution-Share Alike 3. 0 Unported

[2] Coutillot, V. (2003). *Evolutionary Catastrophes: The Science of Mass Extinction.* Cambridge, UK: Cambridge University Press.

Figure 7-14[1] shows the skeleton of an inostrancevia, a member of the therapsid group. These animals lived from 300 to 250 million years ago. Many groups of Therapsids evolved. Many sub-groups of the therapsid group were very large animals, weighing up to a ton. Members in other smaller-sized groups were just the size of a rat.

The Figure 7-15[2] illustrates a cynognathus, a mammal-like animal from the cynodont sub-group. It measures one meter from snout to tail and has a large head measuring thirty centimeters in length.

Figure 7-15 Cynognathus – From the Condant Group – The line becomes more mammal like

Evolution of plants and animals continued until about 250 million years ago. The Permian-Triassic extinction wiped out 90 percent of all species. Not all groups died. This start of the pig-sized plant-eaters called lystrosaurus evolved. As well, large amphibians evolved, who later gave rise to the first true mammals and first dinosaurs.[3]

It took thirty million years to recover from this event. This period gave way to the start of animals that comprise the age of the dinosaurs. These animals and others that evolved predominated the period between 200 million and 65 million years ago. Around 195 million years ago, animals with some similarities to modern-day animals started to appear.

---

[1] https://en. wikipedia. org/wiki/Inostrancevia - fossil of Inostrancevia alexandri by Ghedoghedo is licensed under Creative CommonsAttribution-Share Alike 3. 0 Unported, 2. 5 Generic, 2. 0 Generic and 1. 0 Generic license

[2] https://en. wikipedia. org/wiki/Cynognathus - Cynognathus crateronotus by Nobu Tamura (http://spinops. blogspot. com) is licensed under Creative CommonsCreative CommonsAttribution-Share Alike 4. 0 International license

[3] V. Coutillot. Evolutionary Catastrophes: The Science of Mass Extinction. Cambridge, UK: Cambridge University Press, 2003.

The Age of the Dinosaurs

Figure 7-16 The Proterosuchid – A member of the Archosauriformes – The ruling class, opening the age of dinosaurs

Archosauriformes, a previously obscure snapsid, whose name means "ruling lizards," became the dominant species during the beginning of the recovery from the Permian-Triassic Extinction Event.[1]

As figure 7-16 shows, Archosauriformes bear a great similarity to modern day alligators. Scientists say that these animals appeared as early as 250 million years ago, before the Permian-Jurassic Extinction Event was over.

Archosaurs replaced the archosauriformes about 230 million years ago, but dinosaurs overtook them, about 210 million years ago. Dinosaurs evolved into many types. Figure 7-17[2] shows depictions of some of them.

Approximately 160 million years ago, ancient birds evolved from small predatory theropod dinosaurs. But this group became extinct as well.

It appears that not a single species of dinosaurs survived. Mammals thrived after

Figure 7-17- Various depictions of dinosaurs

---

[1] https://en. wikipedia. org/wiki/Archosauriformes- Proterosuchus fergusi by ДиБгд at Russian WikipediaThis work has been released into the public domain by its author, ДиБгд

[2] https://en. wikipedia. org/wiki/Sauropoda- Camarasaurus, Brachiosaurus, Giraffatitan, Euhelopus by Богданов dmitrchel@mail. ru is licensed under Creative CommonsAttribution-Share Alike 3. 0 Unported license

the end of the dinosaurs. A question exists whether the extinction is related to the increase of mammals, or if the demise of dinosaurs created circumstance more favourable for large mammals.[1]

Mammals: The Door for Manlike Animals

Today about 4,600 species of animals form the mammal group. Unlike other groups, they all have similar characteristics, such as mammary glands, single lower jaw bones, and a neocortex in the forebrain. This "mammalia" group is remarkably distinct from their closest living relatives.[2]

Mammals evolved from early animals. During the age of the dinosaurs, mammals were very small and insignificant animals. Mammals evolved into more and more significant animals. The evolutionary chain evolved the primate species, from where man evolved.

Figure 7-18 Modern Giant Ant Eater – Descendent of nocturnal mammals that survived the age of dinosaurs

---

[1] T. Kemp. The Origin and Evolution of Mammals. New York, NY, USA: Oxford University Press, 2005.
[2] T. Kemp. The Origin and Evolution of Mammals. New York, NY, USA: Oxford University Press, 2005.

Two groups of animals resulting from the early cynodonts managed to survive the age of the dinosaur.[1] These were the tritylodonts and the mammals.

Nature relegated most surviving mammals of this age to small nocturnal insectivores or insect eaters. As well, survivors included a small number of carnivores that fed on vertebrate prey and herbivores that fed on plants. The giant anteater is one modern example of such a mammal, shown in figure 7-18.[2]

These animals probably ushered the way for modern mammals. By 195 million years ago there were animals that were very like today's mammals in many respects. Unfortunately, there is a gap in the fossil record throughout the middle Jurassic.

Dinosaurs, along with many other organisms, perished sixty-six million years ago in the Cretaceous-Paleogene extinction event. The numbers of mammal species were very limited during this period, but the extinction event ushered in a rapid increase in mammals.

One of the signature characteristics of mammals[3] are mammary glands on the female. Mammary glands produce milk to feed newly born offspring. Early Triassic therapsids had two bones at the back of their upper and lower jawbones. In mammals, these bones adapted for use in hearing.

Thirty million years ago, the age of the mammal arrived. Some scientists defined a group called mammaliaforms. Mammaliaforms split into two groups, one being the crown group of mammals. Modern mammals evolved from the crown group of mammals.[4]

---

[1] https://en. wikipedia. org/wiki/Evolution_of_mammals
[2] https://en. wikipedia. org/wiki/Giant_anteater- Myrmecophaga tridactyla by Malene Thyssen is licensed under Creative CommonsAttribution-Share Alike 3. 0 Unported license
[3] https://en. wikipedia. org/wiki/Evolution_of_mammals
[4] Coutillot, V. (2003). Evolutionary Catastrophes: The Science of Mass Extinction. Cambridge, UK: Cambridge University Press.

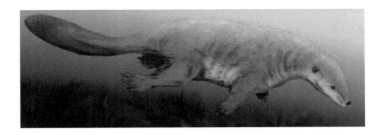

Figure 7-19 - Castorocauda– Early member of the Crown Group of mammals, the group man descended from

Figure 7-19[1] shows the Castorocauda, one small animal in the crown group of animals.

The crown group of mammals evolved into many subgroups. Two of these subgroups make up many of the mammals we see today. One group are marsupials that carry their young in pouches. The other group, placentals, carry their fetus in the uterus or womb of the mother, where the placenta nourishes it.

Scientists have classified mammals into groups called orders.[2] I discuss a few of the well-known orders here. The Afrosoricida order contains animals such as shrews, hedgehogs and moles. The Artiodactyla order contains hoofed animals such as the pig, hippopotamus, deer, caribou, cattle, buffalo, antelope, and giraffe. The Carnivora order contains animals that eat meat as well as animals with meat and plant diets. Examples of the Carnivora order are the cat, otter, hyena, mongoose, dog, bear, badger, skunk, weasel, wolverine, mink, walrus, and the seal. The Cetacea order contains marine animals such as the whale, dolphin, and porpoise.

---

[1] https://en. wikipedia. org/wiki/Castorocauda- Castorocauda lutrasimilis by Nobu Tamura (http://spinops. blogspot. com is licensed under Creative CommonsAttribution-Share Alike 3. 0 Unported license

[2] https://species.wikimedia.org/wiki/SahelanthropusSahelanthropus, an extinct hominid species, possibly a human ancestor, by Sisyphos23, Own work, based mostly on Sahelanthropustchadensis ZICA created by Mateus Zica is licensed under the Creative Commons Attribution-Share Alike 3.0 Unported license.

The Primate Order – Where Man Came From

Figure 7-20 Plesiadapis – One of the first primates

Another important mammal order is the primate order. Man descended from the primate order. Genetic studies show that primates diverged from other mammals, eighty-five million years ago. The earliest fossils appear fifty-five million years ago. Figure 7-20[1] shows the Plesiadapis, one of the oldest primate like animals known. It existed fifty-five to fifty-eight million years ago.

Evidence of the progression of later primate animals is scarce. Scientists believe that fossils found in Kenya and Greece might be the ancestors of modern-day gorillas and, later, chimpanzees. Some scientists think that the gorillas and then the chimpanzees split from the lineage of their common mammal ancestor some four to seven million years ago. Some scientists think that chimpanzees are the ancestors of the human species. Gorillas and chimpanzees have DNA approximately 98.4 percent identical to human DNA. Other scientists think that the lineage of humans and chimpanzees is parallel. In the parallel lineage theory, the lineage of both humans and chimpanzees may have split from a common earlier ancestor. In this case, humans are descendants of the gorilla directly. In any case, scientists appear unanimous that humans evolved from either gorillas or chimpanzees. (Beard 2004) Various species of have similar DNA. The similarities of monkey, ape,

---

[1] https://en. wikipedia. org/wiki/Plesiadapis- *Plesiadapis* by Nobu Tamura (http://spinops. blogspot. com is licensed under Creative CommonsAttribution-Share Alike 3. 0 Unported license

and human DNA is stronger than the similarities of all members of this group to other earlier primate groups.[1]

Figure 7-21 Sahelanthropus – An ape believed to be man's ancestor

Figure 7-22 Modern Western Gorilla - Man possibly descended from this group

Robert Chambers in 1844 and Charles Darwin in 1871 proposed that humans evolved from upright-walking apes called Sahelanthropus. These apes lived approximately seven million years ago. The reconstruction in figure 7-21[2] shows that the Sahelanthropus has a crude resemblance to modern day humans.

Most scientists agree, that humans originated between five and seven million years ago in Africa, and that early humans started to walk upright millions of years before their brains enlarged much beyond those of chimpanzees.

The chimpanzee-sized brains of the earliest hominines appear to have undergone a fourfold increase in size by three million years ago. The first known stone tools are approximately 2.5 million years old. There is a vigorous debate regarding the evolution of the human species. One

---

[1] Beard, C. (2004). The Hunt for the Dawn Monkey. Berkeley and Los Angeles, California, USA: University of California Press.

[2] *https://species.wikimedia.org/wiki/SahelanthropusSahelanthropus*, an extinct hominid species, possibly a human ancestor, by Sisyphos23, Own work, based mostly on SahelanthropustchadensisZICA created by Mateus Zica is licensed under the Creative Commons Attribution-Share Alike 3.0 Unported license.

argument states that humans evolved all over the world from existing hominines. Another argument postulates that humans evolved from a single small population in Africa[1] and then migrated over the world about two hundred thousand years ago. Hominines include a small group of primates, which includes humans, gorillas, and chimpanzees. Figure 7-22[2] shows a Western Gorilla, which is part of this group.

The evolution of hominid family, or ape family, resulted in many classifications of animals along the evolutionary chain leading to humans. One such animal was the Australopithecus afarensis, which in Latin means "southern ape from afar."[3] Figure 7-23 shows a reconstruction of the skeleton of Lucy. Lucy is a famous ape, due to a discovery of her partial skeleton, by Donald Johanson and his colleagues. This ape lived in eastern Africa about 3.2 million years ago but became extinct about two million years ago. Apes were knuckle walkers. Notice how humanlike Lucy's skeleton is. Modern work on the Afarensis group, to reevaluate previous assertions that its intrinsic hand proportions were similar to modern humans confirms that this group is a common link between apes and man. Data shows that the proportions of early human fossils overlap those of the gorilla, approximately with a 95 percent confidence level.[4]

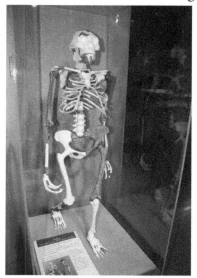

Figure 7-23 - Skeletal reconstruction of Lucy, an Australopithecus Afarensis, a part of the chain leading to humans

---

[1] https://en. wikipedia. org/wiki/Homininae
[2] https://en. wikipedia. org/wiki/Western_gorilla - Gorilla gorilla by Macinate is licensed under Creative CommonsAttribution 2. 0 Generic license
[3] https://en. wikipedia. org/wiki/Lucy_%28Australopithecus%29- Lucy - Cleveland Natural History Museum by Andrew from Cleveland, Ohio, USA is licensed under Creative CommonsAttribution 2. 0 Generic license
[4] Campbell, R. a. (2013). *Reassessing Manual Proportions in Australopithecus.* Calgary, Alberta, Canada: University of Calgary - AMERICAN JOURNAL

Man Emerges

Figure 7-24
Australopithecus afarensis – A predecessor to the Homo sapien group

Man evolved from the primate group of animals approximately five to seven million years ago. Early man was a very primitive and a more animal like being. Gradually man developed into the Homo sapien species we have today. The Hominid species increased rapidly from about 4.6 million years ago but started to decline after two million years ago due to increasing climate instability.[1]

Australopithecus is a group of apes leading to Homo sapiens or humans that lived four million years ago. Figure 7-24[2] shows a reconstruction of Australopithecus Afarensis, a member of this group. The group became extinct about two million years ago. Australopithecus was one of the first hominids to show the presence of a gene that causes increased length and increased ability of neurons in the brain.

Australopithecus played an important role in the evolution of humans. This group lead to the Genus Homo group about three million years ago. About two million years ago, descendants from this group formed groups such as Homo habilis, which eventually led to Homo sapiens or modern humans. Figure 7-25[3] shows Homo habilis, which started a trend of increasing brain size. Fast brain growth characterizes human evolution, beginning with

---

OF PHYSICAL ANTHROPOLOGY.
[1] Finlayson, C. (2004). *Neanderthals and Modern Humans.* Cambridge University Press.
[2] https://en. wikipedia. org/wiki/Australopithecus- Reproducción de Australopithecus afarensis en Cosmocaixa, Barcelona – Author unknown – Public Domain
[3] https://en. wikipedia. org/wiki/Homo_habilis- Homo habilis by Cicero Moraes is licensed under Creative CommonsAttribution-Share Alike 3. 0 Unported license

Homo habilis some two million years ago. This trend continued, resulting in the expansion of Homo erectus and, finally, modern humans.[1] Homo habilis means "handy man" or "able man." Homo habilis used tools for scavenging, such as removing meat from bones.

The size of the homo species brain increased, but his gut size decreased, resulting in the need for a diet with more cooked foods and a more meat-rich diet. Data indicates that the bodies of humans and other primates, developed smaller energy using organs, such as the intestines to make way for the increased energy demands of larger brains. This nescesitated a higher-quality, easy-to-digest diet.

Figure 7-25 Facial reconstruction of Homo habilis – the handy man

Human brain size increased in the temporal lobes, containing centers for language processing and the cerebral cortex, related to decision-making and moderating social behavior. These brain parts also transform past and present experience into future performance. The cerebral cortex in the homo species has developed disproportionately more than other parts of the brain.

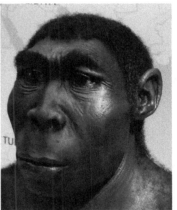

Figure 7-26 Homo erectus, another member in the Homo line

Figure 7-26[2] shows a reconstruction of Homo erectus, a member in the Homo line.[3] Scientists esti-

---

[1] Leiden University Press. (2007). *Guts and Brains*. (W. Roebroeks, Ed.) Leiden University Press.
[2] https://en. wikipedia. org/wiki/Homo_erectus- Homo erectus by Westfälisches Landesmuseum, Herne, Germany is licensed under Creative Commons Attribution-Share Alike 2. 5 Generic, 2. 0 Genericand 1. 0 Genericlicenses
[3] https://en. wikipedia. org/wiki/Homo

mate that Homo erectus appeared in East Africa about 1.9 million years ago. Scientists think that this species may have been the first hunter-gatherers and the first to control fire. Homo erectus became extinct seventy thousand years ago, possibly due to the Toba super eruption catastrophe. Homo erectus spread throughout Africa. One population of Homo erectus left Africa for Eurasia (Java, China, India and Caucasus) 1.3 to 1.8 million years ago.

The name Homo erectus pertains to Asian Homo species, and the name Homo ergaster pertains to African Asian Homo species. Both should probably be sub species of one group.[1]

Figure 7-27[2] **Homo heidelbergensis – Farther down the Homo chain**

Figure 7-27[2] shows Homo heidelbergensis, also known as Homo Afarensis, is an extinct species of the genus Homo that lived in Africa, Europe, and Western Asia between six hundred thousand and two hundred thousand years ago. Some research says that Neanderthals, Denisovans, and modern humans all descended from this species. Other research says that Homo heidelgergenisis is a parallel group that emerged from Homo erectus along with other groups, such as the Neanderthals.[3] Between 300,000 and 400,000 years ago, a group of this species became independent of others, after they left Africa. One group branched into northwest

---

[1] Finlayson, C. (2004). *Neanderthals and Modern Humans.* Cambridge University Press.
[2] https://en. wikipedia. org/wiki/Homo_heidelbergensis- Solo tiene 400. 000 años by Jose Luis Martinez Alvarez from Asturias, España is licensed under Creative CommonsAttribution-Share Alike 2. 0 Genericlicenses
[3] https://en. wikipedia. org/wiki/Human

Europe and West Asia and became the Neanderthals. Another group branched eastward throughout Asia and became the Denisovans.

Homo heidelbergensis may have been the first to bury their dead. Their brain size was 1,100 to 1,400 cm^3, overlapping the average modern human size of 1,350 cm^3. They used crude stone tools. Scientists believe that this species was the first to use a very crude type of vocal communication.

Some scientists treat species from Homo erectus to Homo sapiens as a single group that have produced divergent lineages over the last 1.9 million years. The Neanderthals are the clearest example of such a divergent lineage. Genetic evidence indicates that Neanderthals and modern humans had a common ancestry, around five hundred thousand to four hundred thousand years ago, when the two lineages apparently went along separate paths. Although evidence is scarce, the two lineages probably met again forty thousand years ago.[1]

Scientists propose various models for the expansion of the Homo species. The argument around these varying models questions whether Neanderthals interbred with the Homo species. Whatever the situation, Neanderthals apparently became extinct around this time, leaving no genetic descendants that survived.[2]

Figure 7-28 Neanderthals – Big and rough

Figure 7-28[3] shows a Neanderthal, who was large, including his brain, which occu-

---

[1] C. Finlayson. *Neanderthals and Modern Humans.* Cambridge University Press, 2004.

[2] C. Finlayson. *Neanderthals and Modern Humans.* Cambridge University Press, 2004.

[3] https://en. wikipedia. org/wiki/Neanderthal- Skeleton and restoration model of Neanderthal by Photaro is licensed under Creative CommonsAttribution-Share Alike 3. 0 Unportedlicense

pied 1,600 cm³. The DNA of Neanderthals differs from the DNA of modern humans by less than 0.3 percent. Neanderthals used advanced tools, built dwellings out of animal bones, and travelled on water using dugout canoes. They likely hunted larger animals, such as deer and occasionally a mammoth. There is evidence that they ate cooked vegetables. Evidence shows violence between Neanderthals, as healed skull damage suggests. The consensus is not complete, but scientists believe that Neanderthals interbred with African populations as well as with Homo sapiens or modern humans. The reason for the extinction of Neanderthals is not entirely clear. Climatic change is one possible reason. About fifty-five thousand years ago, weather began to fluctuate wildly and temperatures dropped. This changed the vegetation and food sources of Neanderthals. Densiovians are a primitive manlike species that are genetically distinct from Neanderthals and modern humans.[1] There is evidence that Denisovians inter bred with other species.

Homo sapiens

We are Homo sapiens.[2] The home of the first fossil remains of Homo sapiens is Ethiopia. These remains are two hundred thousand years old. Evidence of Homo sapien fossils in Southern Europe and Israel begins about ninety thousand years ago. One theory postulates that all Homo sapiens started in Africa. A small group left Africa seventy thousand years ago. They reached Eurasia and Oceania by forty thousand years ago and North America as early as 14,500 years ago.

Homo sapiens are the cream of the crop of the homininae group. Until ten thousand years ago, humans lived primarily as hunter-gatherers. These hunter-gatherers lived in small nomadic tribes. Then humans started to live in permanent human settlements. Domestication of animals, agriculture and trade started, and food generally became more plentiful. Metal tools started to make their way into society. We saw previously how early Mesopotamian devel-

---

[1] https://en. wikipedia. org/wiki/Denisovan
[2] https://en. wikipedia. org/wiki/Anatomically_modern_human

opment lead to increasing commerce and trade. Living in groups provides benefits for those in the group.¹

Humans evolved culturally over time and reached the full extent of language, music, and other activities about fifty thousand years ago. The story of Homo sapiens is truly amazing. The two hundred thousand years Homo sapiens were on Earth is only 0.004 percent of the lifespan of the Earth. It took 4.6 billion years for Homo sapiens to form, rising from single-celled bacterialike organisms to entities with human form. For approximately 190,000 years, or 95 percent of their time on Earth, Homo sapiens were wandering hunter gatherers. Then on day 3 of his work, God formed Adam from the dust of the Earth. Man's ways changed to a communal lifestyle. Domestic crops and animals stabilized man's existence. Organized society began to form. Then God imparted his spirit to man through Adam, and civilization and lifestyles exploded forward.

The thought that the LORD in bringing about the universe formed a physical Earth and the things in it is hard to contemplate. It is even harder to contemplate how organic matter formed from a basically inorganic universe. The truly amazing part of this discussion is evolution. How, all on their own, these single-celled organisms evolved over billions of years into thinking, feeling humans is the LORD's most amazing feat of all. The LORD built this ability into his universe when the Big Bang occurred.

Figure 7-29 – Ancient Mesopotamian sickle

---

[1] Leiden University Press. (2007). *Guts and Brains*. (W. Roebroeks, Ed.) Leiden University Press.

The last ten thousand years or so demonstrate the LORD's awesome power. The Earth evolved on its own for billions of years, resulting in a basic man of low grade. Then on day 3 of God's work, the LORD got involved and nudged the process along by forming Adam and domestic animals and plants. What an amazing increase in productivity occurred. And the legacy keeps on growing. The advances man has made during my brief life to date of less than seventy years is astounding. The LORD God's spirit is a growing and lasting legacy.

A look at grain harvesting technology shows this. Figure 7-29[1] shows an ancient Mesopotamian sickle from approximately ten thousand years ago.

Figure 7-30 Modern Sickle

Figure 7-31 A worker using a scythe

Sickles, as shown in figure 7-30,[2] improved with the iron age, but harvesting with a sickle was a backbreaking job, with the harvester bending over to cut and grab cut grain stalks.

Figure 7-31[3] shows a scythe, invented about 2,500 years ago, but first coming into common use five hundred years ago. The scythe, although still a completely manual tool, allowed use in an upright position and increased

---

[1] https://en. wikipedia. org/wiki/Sickle - Neolithic sickle by Wolfgang Sauber is licensed under Creative Commons Attribution-Share Alike 3. 0 Unported, 2. 5 Generic, 2. 0 Generic and 1. 0 Generic licenses
[2] https://en. wikipedia. org/wiki/Sickle- Sickle by Chmee2 is licensed under Creative Creative CommonsAttribution-Share Alike 3. 0 Unportedlicense
[3] https://en. wikipedia. org/wiki/Scythe- Scythe by unknown artist is Public Domain – published in the U. S. before 1923 and public domain in the U. S.

the amount of grain a reaper could cut. Thus, it took about 7,500 years to advance from one archaic form of reaping to another.

Figure 7-32[1] depicts farming practices 3,500 years ago in Egypt. When Adam became the world's first farmer on day 3 of God's work, even the oxen-driven plough, as shown in figure 7-32, did not exist. Even thousands of years later, all aspects of grain agriculture were largely acts of manual labor. Agriculture was a toilsome practice, even in more recent times, as shown in figure 7-33.[2] Workers manually cut down grain, gathered it into bunches, and then tied the bunches into sheaths. Humans then manually threshed the sheaths, on a threshing floor, to separate the grain kernels from the straw and chaff.

Figure 7-32 – Agricultural practices 3,500 years ago as depicted from a picture in the ancient Egyptian Tomb of Nakht

Figure 7-33 – Peter Bruegel's "The Harvesters' of 1565 shows the backbreaking toil of manual harvesting

---

[1] https://en. wikipedia. org/wiki/History_of_agriculture - Agricultural scenesby Norman de Garis Davies is Public Domain – published in the U. S. before 1923 and public domain in the U. S.

[2] https://en. wikipedia. org/wiki/The_Harvesters_%28painting%29- The Harvestersby Peter Bruegel is Public Domain – published in the U. S. before 1923 and public domain in the U. S.

Cyrus McCormick patented the first mechanical reaper in 1831. After over two thousand years of using the scythe, automation started to take hold. Then twenty years later, the first reaper-binder appeared, which not only cut grain down but also tied it into bundles.[1]

I can remember playing with the old binders, used by my grandfather. The use of binders ended just before my birth in 1950. Figure 7-34[2] shows a binder drawn by horses. The binder eliminated the backbreaking work of cutting crops down with manual sickles or scythes and then gathering up the loose grain stalks and tying them into bundles. The binder tied the grain into bundles and then dropped groups of five or so bundles in the field.

Figure 7-34 An early binder cutting grain and tying grain stalks into bundles

This invention increased productivity and decreased the cost of farming significantly. By the 1920s, many farmers used tractors to pull the binders, despite the need for an extra operator, one for the tractor and one for the binder[3]

---

[1] S. Evans. Bound in Twine 1959
[2] https://en. wikipedia. org/wiki/Reaper- Feature. Agricultural Schoolby Conrad Poirier is Public Domain – The Canadian copyright expired
[3] S. Evans. Bound in Twine, 1959.

Figure 7-35 – Sheaves or "bundles' of grain drying after workers place them in stooks

Figure 7-36 – A threshing machine driven by belt with a tractor and with workers using forks to load bundles from horse drawn racks into the threshing machine

But the binder did not eliminate the hard work completely. Workers still had to stack the gathered sheaves of grain into stooks, where they dried, thus maturing the grain for threshing. Figure 7-35[1] shows stooks drying in the field.

Along with the binder, came the threshing machine, which eliminated another toilsome job, threshing grain on a threshing floor. Instead of loading and hauling the sheaves or bundles of grain to threshing floor, workers hauled them to a threshing machine. A steam or gasoline tractor generally powered the threshing machine. The operation still required considerable manual labor to load the bundles onto horse drawn racks, haul them to the threshing machine, and then manually unload them into the threshing machine. The threshing machine separated the grain from the straw and chaff. Figure 7-36[2] shows a threshing machine.

---

[1] https://en. wikipedia. org/wiki/Stook- Wheat sheaves near King's Somborne by Trish Steel (Trish Steel) is licensed under Creative CommonsAttribution-Share Alike 2. 0 Generic license

[2] https://en. wikipedia. org/wiki/Threshing_machine - Threshing machine in action by Ben Franske is licensed under Creative Commons Attribution-Share Alike 4. 0 International, 3. 0 Unported, 2. 5 Generic, 2. 0 Generic and 1. 0 Generic licenses

Figure 7-37 – an early model combine

Figure 7-38 – a newer model combine

The combine combined all operations into one. One machine cut and threshed the grain in one operation, requiring only one operator. Farmers often employed one other worker to haul grain from the combine to bins. The combine shown Figure 7-38 uses a pick up to harvest grain previously cut and placed into swaths. Farmers use this operation to mature grain faster by cutting it down first for drying. Combines relegated most binders and threshing machines to museums by 1950.

There were 36,734 combines on Canadian Prairie Province farms by 1946. Data from 1950 showed a steady increase in both self-propelled and tractor-pulled combines. The figures perplexed twine manufacturing.[1] Figure 7-37 shows an early combine that my dad owned around the year 1950. Figure 7-38 shows a combine my Dad owned prior to the 1990s.

It took 4.5 billion years for man to develop as hunter-gatherers. Then about ten thousand years ago, man started to harvest wild grains, albeit the hard way: by hand. Eighty years ago, man invented the first binder, eliminating much of the toilsome work and increasing productivity manifold. Sixty years ago, combines replaced the binder and threshing machine. I used to operate combines for my father before I left for university in 1968. Now another fifty years later, my cousins who remained as farmers told me I could not be a farmer with my past knowledge of farming. Things

---

[1] S. Evans. *Bound in Twine, 1959.*

have again changed so much that my past knowledge is obsolete, as the combine in figure 7-39[1] shows.

Figure 7-39 – Modern state of the art combine

The acceleration of development in farming and virtually every other area shows what the LORD God's Spirit can do. We have come farther faster than any predecessor before us. Our knowledge compounds upon itself as time progresses. We are the only known organism in the universe where this is possible. So why are we so special; the answer is God's spirit.

---

[1] https://en. wikipedia. org/wiki/Combine_harvester- Threshing machine in action by Kowloonese is licensed under Creative CommonsAttribution 3. 0 Unported license

8

Formation of the Universe and Earth

While most of the Bible concentrates on the last fifteen thousand years or so, the Bible also mentions the start of the universe. The Bible mentions how the LORD initiated the Big Bang, starting the whole process of forming the universe and Earth. It then goes on to mention some of the major changes on Earth that formed the oceans, continents, and resulted in the Earth as we see it today. The Earth started off as rugged rock. The Bible mentions the formation of soil that allowed man to have a variety of plants for food, and later animals. These events occurred before recorded history. Early authors wrote the Bible before the advent of science. The only explanation for this knowledge is divine inspiration.

The Big Bang: Physical and Spiritual Formation

J. Baggott[1] feels that some recent popular science books, magazine articles or news features are only trying to fool people when they try to tell us how the universe began. They question even if the word "began" is even appropriate in the context of the beginning of the universe. Mr. Baggott goes on to present theories and the work of contemporary scientists on the matter. He spends two chapters

---

[1] J. Baggott. *Origins - The Scientific Story of Creation.* Oxford, Unite Kingdom: Oxford University Press, 2015.

consisting of sixty pages describing six phases or epochs of the first second after the Big Bang started. The Big Bang starts with temperatures as high as 10^{35} (one with 35 zeroes after it) degrees Kelvin. As the author says, 0.000,000,000,001 seconds after the Big Bang started, subatomic particles formed. By one second after the start of the Big Bang, particles with somewhat familiar names such as electrons, neutrinos, and photons appear. A detailed discussion of this subject is well beyond the scope of this book.

As we will see below, the Bible tells us that the LORD existed before the Big Bang. The Big Bang is totally incomprehensible to most individuals, such as myself. Albert Einstein and scientists like him appear aware of a fourth dimension of existence, outside the three-dimensional existence that most of us perceive. The nature of the LORD before the Big Bang probably requires at least a fifth or sixth dimension of existence, beyond that contemplated by Einstein.

Scientists believe in an initially highly concentrated universe with a very small volume, extremely high density, and extremely high temperature.[1] The Big Bang theory postulates that the universe is expanding; therefore, all parts of the universe must have been closer together in the past. The Big Bang model relies on Albert Einstein's theory of relativity. This theory allows scientists to model the universe back in time. Scientists can model the formation of the universe backward only to a certain point. I will call this "the point beyond knowledge." A considerable problem results at this point. Einstein's relativity theory and all known laws of physics break down. Scientists can prove their theories in laboratory experiments to this point, but not earlier. J. Baggott advises that our present scientific theories do not stand up to describing the beginning of the universe.[2]

The time beyond knowledge lasted for only 10^{-12} seconds, after the start of the formation of the universe. After 10^{-12} seconds, the universe consisted of intense radiation with very little matter. Temperatures during the Big Bang were so high that the term tem-

---

[1] https://en.wikipedia.org/wiki/Big_Bang
[2] J. Baggott. *Origins - The Scientific Story of Creation.* Oxford, Unite Kingdom: Oxford University Press, 2015.

perature loses its meaning. Scientists refer to the pre-matter entities that existed at the time as quark and anti-quark pairs. By 10^{-4} seconds, quarks ceased to exist but combined to make mesons. Some mesons combined to make protons and neutrons, but temperatures were too high to result in atoms. After one second, neutrons started to disintegrate disproportionally to protons, resulting in more protons than neutrons. A minute or two after the start of the Big Bang, the universe cooled to about one billion degrees Celsius and had the density of air and the first atoms formed. Initially, only light atoms such as helium, lithium, and hydrogen formed. Formation into atoms such as hydrogen was not a direct process, and isotopes of hydrogen such as deuterium formed first. It took ten thousand years of cooling to reach a point where matter and energy were independent of one another. After 380,000 years, the universe cooled to 3,000 Kelvin. Hydrogen particles now acquire electrons forming neutral atoms, making way for heavier atoms and combinations of atoms to form compounds. At this point, localized areas of higher density clouds of matter in the universe began to consolidate as gravitational forces drew particles together to form matter. The first stars formed six hundred million years after the Big Bang. The Milky Way formed about ten billion years ago or 3.8 billion years after the Big Bang. The age of the universe is 13.8 billion years.[1][2]

  Several possible new theories are currently under study, regarding the particles present during the formation of the universe. Arthur B. McDonald, director of the Sudbury Neutrino Observatory[3] in Sudbury, Ontario, Canada, received the Nobel Peace Prize in 2015 for his work with neutrinos. Neutrinos[4] are subatomic particles that scientists discovered. One question regarding neutrinos is whether they have mass. Some science seems to suggest that neutrinos have mass when travelling below the speed of light, but above the speed of

---

[1] J. Baggott. *Origins - The Scientific Story of Creation.* Oxford, Unite Kingdom: Oxford University Press, 2015.
[2] M. M. Woolfson. *Time Space Stars and Man.* London, UK: Imperial College Press, 2013.
[3] https://en. wikipedia. org/wiki/Sudbury_Neutrino_Observatory
[4] https://en. wikipedia. org/wiki/Neutrino

light, they become pure energy. (Woolfson 2013) The destruction of a neutron creates a neutrino.

"The point beyond knowledge" brings up an obvious question. Perhaps we must consider that some all-powerful source is responsible for creation. Is an all-powerful, omnipotent source required to explain "the point beyond knowledge"? When speculating about the big bang, scientific theory seems to point toward state where matter ceases to exist as matter and, as implied by Neutrino research, exists as pure energy. Could this be the essence of an all-powerful, omnipotent source—a state of infinite will and knowledge, expressing itself in the creation of matter from energy? Or perhaps the term energy has no meaning prior to the big bang. Perhaps we need some other phenomena, only existent in a dimension not perceivable by man, to even begin to describe the LORD, prior to the Big Bang. The book of Psalms talks about the creation of the universe: "By the word of the LORD the heavens were made, their starry host by the breath of his mouth."[1]

Modern Bibles provide misleading translations of the above passage. The English word "word" does not strictly refer to speech in the sense of the spoken word. Rather than "word" in the contemporary sense, the Hebrew word used connotes action, undertaking, guidance, or resolve. The LORD is a spirit. It is impossible for man to understand the LORD's "action, undertaking, guidance or resolve" in terms of human attributes. Perhaps thought is the closest analogy. However, what is thought?

The Massachusetts Institute of Technology, in an Internet posting,[2] essentially defines thought as electrical energy travelling through our brains. Electrical signals propagate along long threads called axons. Synapses at the end of axons retransmit these electrical signals or thoughts to other axons. Therefore, thoughts then are simply energy.

The LORD's word obviously is more than just speech. The LORD's word causes action and even causes matter to form. The matter that formed was the "heavens." The original Hebrew text talks about the LORD's will and resolve that turned into the universe.

---

[1] Bible, Book of Psalms, Ch. 33, Vs 6 & 7
[2] http://engineering. mit. edu/ask/what-are-thoughts-made

Notice that the Bible does not use the word "create" for the formation of the universe. The original Bible uses a Hebrew word for the English words "were made" that means an act of achievement or moving things forward. Thus, through his resolve, the LORD achieved the universe.

The LORD not only brought forth the universe but also an army of spiritual beings as part of the achievement of the universe. Modern translations of the Bible lose this point. Modern Bibles interpret the last part of the above passage from Psalms as "their starry host by the breath of his mouth." This is misleading. Unlike modern translations, the original Bible does not include the word "starry." Modern Bibles imply that these words talk about the formation of stars, but this is not the case.

The original Bible uses a Hebrew word for "host" that means a legion or army, with the connotation of a confrontation or a campaign. I call this large army of beings the "heavenly host." Along with the universe, the LORD brought forth a heavenly host of spiritual beings. The Bible thus tells us that the LORD in bringing about the universe did not form only matter. Some of the energy turned into spiritual beings. These beings possessed qualities more like the LORD's qualities than human qualities.

The original Bible text uses two words for English words "by the breath of his mouth." The first Hebrew word used for the English words "by the breath" means a sudden and furious action, of a mindful, reasonable spirit; also wrath or "fury." Psalms captures the intensity and heat that science tells us was part of the Big Bang when forming the heavenly host.

The second Hebrew word used for the English words "of his mouth" does not refer to a mouth in a physical sense. The Hebrew word means an action, undertaking, guidance, or resolve, The Hebrew word comes from a root word that has connotations of a strong exhalation that disperses to faraway places or to all locations. This passage tells us that the LORD dispersed the heavenly host to all parts of the vast universe. This is an important point. As we discussed previously, some of the heavenly host noticed that the Earth was the gem of the universe.

The Timelessness and Infinite Size of the Universe

Modern versions of the Bible use the word "heaven" or "heavens" in many places. The original Hebrew Bible uses many different words for the English word "heaven" or "heavens." The original Hebrew words used for "heavens" can mean anything from the sky immediately surrounding Earth to the totality of the universe or outer space. In the passage above, the Bible uses a Hebrew word for the English word "heavens" that encompasses the totality of the universe. Another definition for the word is the place where the LORD resides.

Scientists believe that the solar system formed from a cloud of matter sixty-five light-years across[1] in size. The cloud coalesced into fragments called pre-solar nebula, roughly three to four light years in size. Farther collapse of one of the fragments into dense bodies, about 0.4 light years in size, eventually became our solar system. A light-year is the distance light travels in one year. Light travels at approximately three hundred thousand kilometers per second.[2] Sixty-five light years is 614×10^{12} kilometers. This distance is like travelling around the Earth fifteen billion times. The solar system is one of many such systems in our home galaxy, the MilkyWay Galaxy. There are very many galaxies in space. Obviously, the size of space is immense and beyond the comprehension of most people.

Figure 8-1 Distance from sun to Earth

Figure 8-1 shows the distance between the sun and the Earth. The illustration is to scale, and the size of the sun appears relative to the 149,600,000-kilometer distance between the sun and Earth. When drawn to scale, the sun appears as a large dot, even though

---

[1] https://en. wikipedia. org/wiki/Formation_and_evolution_of_the_Solar_System
[2] https://en. wikipedia. org/wiki/Speed_of_light

it is 865,000 kilometers in diameter.[1] The Earth, which is 12,756 kilometers in diameter, appears just as a small, barely discernable dot. Light travels at three hundred thousand kilometers per second, so the Earth is 0.000096 light-years from the sun. It is mind-boggling to contemplate the distance to faraway galaxies. The distance from the sun to the Earth is very large in human terms, but this distance is only an infinitesimally small fraction of the distance to the nearest galaxies. The word used in Psalms for the English word "heavens" is an apt description of what we know about the size of the universe today.

The Book of Proverbs

Wisdom is a spiritual entity that we read about in the book of Proverbs. "The LORD brought me forth as the first of his works, before his deeds of old. I was formed long ages ago, at the very beginning, when the world came to be."[2] The wording of this passage tells us that the Bible speaks about the earliest possible time, even before the Big Bang. It also tells us that before the LORD brought forth the universe and the heavenly host, he brought forth Wisdom.

The original Hebrew text does not use the phrase "brought me forth" but rather a word that means "owned me." Thus, it seems that Wisdom was not a created entity, but rather a quality the LORD possessed or owned. The LORD brought wisdom forth from himself or poured wisdom out as one of his first works. Thus, wisdom is the very first of the spiritual beings emanating from the LORD.

The original Bible uses Hebrew words for "before his deeds of old" that indicates a spectacular accomplishment. This wording undoubtedly refers to the formation of the universe. As many as four hundred billion stars and other bodies exist in the Milky Way Galaxy.[3] Astronomers know of 125 billion. What accomplishment is more spectacular than bringing forth something as

---

[1] https://en. wikipedia. org/wiki/Sun
[2] Proverbs 8:22–23
[3] D. J. Eicher. *Comets - Visitors From Deeppace.* New York: Cambridge University Press, 2013.

vast as the universe? Even more spectacular than its vastness, is the idea that the LORD brought forth the universe from himself. As shown in the previous section the universe came about by "by the word of the LORD." Thus, in some unimaginable way, the universe always existed. It was part of the LORD. The LORD only had to let the universe evolve from himself by some process no mere human can even remotely grasp.

Thus, the Bible talks about a time that precedes the Big Bang. We could possibly characterize this point as the beginning of time. The wording implies the earliest time possible and almost certainly places the time at or before the big bang. Wisdom was there before the creation of the universe.

The Solar System and Earth Appear

Scientific theory has the solar system as a giant disc at its formation.[1] Gravity from the disc pulled matter inward. Jim Baggott suggests that pictures obtained by the Hubble Space Telescope of the Eagle Nebula forming in the Constellation Serpens provide a picture of what the solar system looked like, at the beginning of its formation.[2] The author indicates that the clouds forming the Eagle Nebula are several light-years long. See figure 8-2.[3]

Figure 8-2 – Eagle Nebula forming in the constellation Serpens provides a picture of what the formation of the Solar System may have looked like

---

[1] D. J. Eicher. *Comets - Visitors From Deeppace*. New York: Cambridge University Press, 2013.
[2] J. Baggott. *Origins - The Scientific Story of Creation*. Oxford, Unite Kingdom: Oxford University Press, 2015.
[3] https://en. wikipedia. org/wiki/Pillars_of_Creation- Star forming pillars in the Eagle Nebulaby NASA, Jeff Hester, and Paul Scowen (Arizona State University)

Very simply the Earth began as a mixture consisting mainly of iron and silicates. Gravity drew the much heavier iron to the Earth's core. The silicates floated to the top of the mixture to form the Earth's lower mantle. The core formed over a roughly thirty-eight-million-year time span and the lower mantle took another thirty-seven million years to form.

Modern science tells us that hot gases from the Big Bang eventually formed the universe and the Earth. The Earth was initially in a very hot molten state. The Earth had no water at this point because the temperature was above that at which water could survive. The Bible eludes to this hot and waterless condition by telling us that wisdom was there before water arrived. "When there were no watery depths, I was given birth, when there were no springs overflowing with water."[1]

The Bible also talks about a time after the Big Bang. "I was there when he set the heavens in place."[2] The original Bible uses a Hebrew word for heavens that eludes to the heavens as someone viewing the entirety of the heavens from Earth. As we look up at night, we see the stars, moon, sun, and some of the planets. We also sometimes see comets and asteroids that venture into the Earth's atmosphere. In this verse, the Bible talks about a time early in the history of the universe, before the stars and planets were in the locations we see them today.

The original Bible uses a Hebrew word for the English word "set" that implies a settling, a making secure, and a stabilization. The heavens looked much different during the Earth's formation than today. After their original formation, asteroids and comets passed by the Earth, and some crashed to the Earth. The planets also continuously changed position during this time and were significantly closer to Earth. Some scientists[3] tell us that the planets were not in the present order from the sun. Some planets had to exchange orbits to result in the solar system as it is today. The universe was a young "expand-

---

is in the public domain because it was created by NASA and ESA
[1] Proverbs 8:24
[2] Proverbs 8:27–28
[3] J. Baggott. *Origins - The Scientific Story of Creation*. Oxford, Unite Kingdom: Oxford University Press, 2015.

ing universe," and wisdom noticed the changes in position of nearby galaxies. Over time, wisdom could see the changes in the position of nearby planets. All the planets were closer to the sun during this time and were moving away from the sun. These changes made the skies look much different as time progressed. Space is expanding and moving outward. Galaxies were much closer together three to four billion years ago. As galaxies moved outward in space, changes in the night sky occurred over time. Wisdom noticed the changes in the position of the stars over time.

Today we look at the night sky. The number and position of the stars seemingly never changes, over the years. In reality, the stars always change position, as galaxies drift farther and farther into space. Today, the distance of the stars to Earth is so large, that we cannot see the changes from Earth. Wisdom was there to witness this when the LORD "set the heavens in place," resulting in the stable heavens we see today.

The sun and solar system formed 4.57 billion years ago. Science[1] hypothesizes that the planets formed from circulating dust left over from the formation of the sun. The dust orbiting the sun formed into clumps of ever-increasing diameter. By approximately one million years, these clumps collided to form bodies up to a kilometer in size. Continuing collisions occurring over the next few million years, forming larger and larger bodies, resulting in the planets we know today. The Earth collided with many large bodies, even as large as the planet Mars. Scientists speculate that the composition of the Earth's layers varied with time. Initially high melting point or "refractory" materials formed the Earth's inner layers. As time went on, temperature dropped and lower, melting point or volatile materials accumulated in the Earth's outer layers. Any lighter compounds that arrived with heavier compounds bubbled up to the Earth's surface, creating huge and violent convective currents. The Earth could only form as we know it today if the composition of the clumps varied during the various stages of the Earth's formation.

---

[1] K. C. Condie. *Earth as an Evolvong Planetary System*. Oxford, UK: Acedemic Press, 2011.

Formation of the Deep

Scientists have recently discovered the transition zone, or as the Bible calls it, the "deep."[1] The deep refers to a large water-bearing layer of the Earth's upper mantle between 410 and 660 kilometers beneath the Earth's surface. Recent discovery of a special diamond by scientists at the University of Alberta[2] has sparked renewed interest in the subject. Scientists long thought of the transition zone between the upper and lower mantle as relatively dry. However, Graham Pearson of the University of Alberta advises otherwise. The University of Alberta scientists discovered Ringwoodite inside of a diamond found in Brazil. Ringwoodite is a form of the green mineral Olivine. The grain, in the diamond showed evidence of water. The scientists determined that a volcano discharged the diamond from four hundred to six hundred kilometers below the Earth's surface to the Earth's surface.

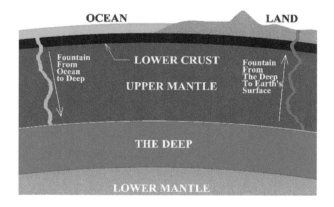

Figure 8-3 - Fountains connect the deep to the Earth's surface

---

[1] American Geophysical Union. *Earth's Deep Water Cycle*. Washington, DC, Usa: American Geophysical Union, 2006.
[2] This section generally taken from Wikipedia http://beginningandend.com/scientists-confirm-biblical-account-of-the-fountains-of-the-deep/

The University of Alberta scientists believe that the Earth recycles water from the Earth's surface and oceans to and from the "deep" through tectonic plates that were originally at the bottom of the oceans. Scientists only recently acknowledged that water might be recirculating in this way. According to the article referenced below, Mr. Pearson feels that the water lost to the Earth's surface by geysers, volcanoes, and the like must recycle back underground. Otherwise, the interior of the Earth would just become a drier and drier. The water, now known to exist deep in the Earth, may help scientists better understand major geological processes. Mr. Pearson advised that water lowers the melting point of rock, so water dictates the location of many of the Earth's volcanoes.

Figure 8-3 depicts the deep and the fountains of the deep that connect the deep to the Earth's surface. Discontinuities in tectonic plates likely form part of the fountains of the deep. For clarity, figure 8-3 shows the ocean, land, and the upper crust ten times thicker than is the actual case.

Outgassing from the Earth's interior probably formed some volatile compounds.[1] But the source of many of the volatile compounds, especially water, probably was not near the vicinity of the forming Earth, with temperatures too high to condense water. Water-rich comets from the outer solar system delivered water as well as compounds such as organics, carbon dioxide, carbon monoxide, and ammonia. Their collision with the Earth resulted in an explosion of flying rock and water. Much of the water was in vapour form, as the incoming asteroids and/or comets heated up in the Earth's atmosphere. The incoming rock and water landed on the still hot surface of the lower mantle, causing yet more

Figure 8-4 – Comets build-up the deep or transition zone

---

[1] J. I. Lunine. *Earth - Evolution of a Habitable World.* New York, NY, USA: CAMBRIDGE UNIVERSITY PRESS, 1999.

vapour and steam. Figure 8-4 depicts the Earth at the time. It was a formless surging mass of rock, water, and steam.

The Bible uses a Hebrew word for "deep" that means a subterranean, netherworld, without shape and where chaos and disorder reign; in other words, "hell on Earth." The deep is at an extremely high temperature and pressure. The Bible when referring to the deep refers to the transition zone.

The Earth's upper mantle formed to cover up the deep after the deep's formation. The formation of the upper mantle covered up the boiling, surging deep. The surface of the Earth, previously obscure and hard to see, now took form. Wisdom was there to witness this "when he marked out the horizon on the face of the deep."[1]

The original Hebrew word used for the English words "when he marked out" means to assign, declare, or mandate, with the connotation of ordaining by carving in rock or metal. What an apt description of what happened. Solid rock covered the deep.

The Hebrew word for the English word "edge" means something round or in the shape of an arc. The surface of the Earth now started to look spherical in shape. Figure 8-5 depicts what the Bible says the Earth's may have looked like.

The Hebrew word for "over" means "on top of" or "counter to." It has a connotation of pushing downward. The Hebrew word for "the face" implies a countenance of combat, conflict, or controversy. The Bible tells of how the overburden of rock covering the deep was in opposition to the deep and covered up the angry face of the deep. The Bible presents an apt description of what science tells us about the deep and how it formed.

Figure 8-5 – Earth's surface after asteroid and comet bombardment

---

[1] Proverbs 8:27–28

(Condie 2011) Scientists[1] have various theories for the origin of the Earth's crust. The impact theory postulates that solid objects such as asteroids and comets struck the Earth. Other theories propose that the process of accumulating clumps added a thin final veneer to the Earth's surface. Many scientists seem to adhere to a late bombardment theory, where a storm of asteroids and comets bombarded the Earth, during its later stages of formation. The asteroid belt exists at the edge of the terrestrial region, between Mars and Jupiter. Initially the asteroid belt had enough matter to form two to three Earth-like planets. Gravitational interactions between Jupiter and the asteroids swept some of the asteroids away from the asteroid belt. Other asteroids migrated towards the center of the solar system, where they collided with the inner planets. Asteroids swept away from the asteroid belt and asteroids colliding with the terrestrial planets reduced the mass of the asteroid belt to less than one per cent of the mass of the Earth. The mass of the asteroid belt later depleted to approximately one fifth of one per cent the mass of the Earth, due to orbital changes of the planets Jupiter and Saturn. Scientists postulate that Earth's moon is the result of a collision of a body with Earth that displaced part of the Earth's mantle into space. This debris formed the Moon we see today.

---

[1] K. C. Condie. *Earth as an Evolvong Planetary System.* Oxford, UK: Acedemic Press, 2011.

Land and Then Oceans

Figure 8-6 - The Earth after the comet and asteroid bombardment – Water has not reached the Earth's surface yet.

After the initial formation of the crust, the Earth's surface was a rocky, barren place. It is unclear if water existed on the Earth's surface or if water trapped in the deep rose to water the Earth. If we allow the Bible to guide us in this question, a dry Earth scenario seems likely. We do know that initially the Earth had no continents and land consisted of many small separated land masses.

Some refer to the first land on Earth as Vaalbara. Figure 8-6 depicts the Earth's early surface. No water has yet appeared near land.

As the Earth's upper mantle formed, it exerted huge pressure on the deep. Eventually passages between the deep and the Earth's surface developed. Water from the deep started to move from the deep to the Earth's surface. The water in the deep was hot and so when it arrived at the Earth's surface, part of it readily evaporated to form fog and clouds.

Wisdom talks about a time when no water existed on Earth; "When there were no watery depths, I was given birth, when there were no springs overflowing with water."[1] The original Hebrew text

---

[1] Proverbs 8:24

uses a Hebrew word for "watery depths" refers to the "oceans." Thus, the Earth's surface was relatively if not completely dry.

The Hebrew word for "springs" means a spring or well, with a connotation of satisfaction. The original Hebrew word used for "overflowing" implies a bourgeoning flow rate with a connotation of being plentiful and bountiful. Thus, springs brought water from the deep to form the Earth's oceans.

Proverbs tells us about how water arriving on the Earth's surface formed clouds; "when he established the clouds above."[1] The original Bible text uses a word for "when he established" that implies a fearless and unwavering persistence. The Hebrew word for "the clouds," does not necessarily mean clouds, but could mean the skies in general. The word could also mean powder. Perhaps the last small remnants of dust from the comets had not all fallen to Earth yet. This wording is amazing, given the results of recent sampling of dust from comets. The Deep Impact spacecraft impacted Comet 103 PHartley 2, releasing dust from the comet's nucleus. Released material was fine-grained, more like talcum powder rather than sand.[2]

Water leaving the deep needs to return to the deep; otherwise, as the scientists at the university of Alberta tell us, the Earth would dry up. The Bible tells us that the LORD included this consideration in his work "and fixed securely the fountains of the deep."[3]

The original Bible text uses a word for "fixed securely" that could imply "overcame." The word also implies he toughened or reinforced the springs of the deep. The Hebrew word for the word "fountains" means "eyes" or "lenses." The word also implies the acceptability of the external appearance of the eyes of the deep. Here God's work seems concerned with assuring the sturdiness of the fountains, connecting the deep with the Earth's surface. It seems that God may have changed or beautified the appearance of the openings.

---

[1] Proverbs 8:28
[2] D. J. Eicher. *Comets - Visitors From Deeppace.* New York: Cambridge University Press, 2013.
[3] Proverbs 8:28

**Figure 8-7 Excess water squeezed up from
the deep starts to appear near land**

After the oceans started to fill up with water from the deep, water started to appear near land. Figure 8-7 depicts the scene after the springs of the deep supplied enough to the Earth's surface to make water visible near land. The water was hot and so the atmosphere takes on a foggy, steamy look. Water is just starting to near the level of normal sea level.

Figure 8-8 shows the ocean near its normal level. Now the fog is so thick that only nearby land and the ocean are visible. Hot water rising up from the deep forms fog and clouds.

Again, the Bible captures the essence of what science tells us about the formation of the Earth.

**Figure 8-8 Water reaches normal ocean levels
– Water temperature creates fog**

Continents Form – Oceans Deepen

The first land, Vaalbara, began as many scattered separate landmasses. These separate landmasses consolidated into larger continents because of upheaval of the ocean floor. The newly shaped continents changed the face of the Earth. Figure 8-9 depicts a scene that replaces the previously scattered land masses.

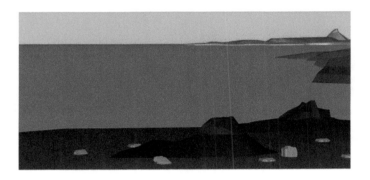

Figure 8-9 – Larger land masses form, the ocean consolidates

The Earth endured many continental building cycles. Wisdom witnessed this process: "When he gave the sea its boundary so the waters would not overstep his command."[1] Larger landmasses now prevail beside the oceans. The ocean now is not a constant predator of the land by completely and closely encircling small, isolated landmasses.

The Earth did not have hills or mountains before the upheaval of the sea-formed continents. Science tells us that the formation of continents pushed up large mountains far above the Earth's surface. As well, smaller hills formed. However, continents and mountains are heavy. They would sink back into the Earth, if not for large foundations. The LORD provided for this: "When he marked out the foundations of the Earth."

---

[1] Proverbs 8:29

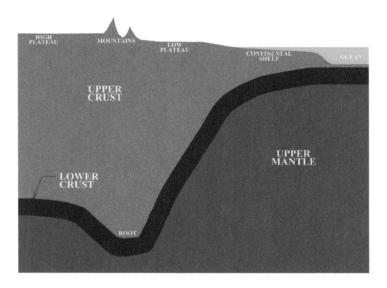

**Figure 8-10 - Land needs foundations to support
it's weight at elevations above the oceans**

The Hebrew word for "marked" is the same word describing how rock covered the deep. The word means to inscribe, with the connotation of carving on rock, and is particularly appropriate. The Hebrew word used for Earth refers to the Earth in an overall sense, referring to the geography of the Earth. As the continents moved around, the geography of the world changed.

The original Hebrew text provides a very good sense of the nature of the changes. Lighter and less dense rocks make up the continents compared to the mantle on which they rest. Continental rocks are lighter, and they float on top of the denser basic rocks below. The oceans have a very thin layer of lighter rock under them compared to the continents. Water is less dense than rock and oceans do not need a deep foundation. Thus, continents protrude upwards, higher than the oceans.

As shown in figure 8-10, the small difference in density of the rock layers results in huge foundations under land masses, especially mountains. This foundation design of the Earth is an absolute necessity to prevent the continents from sinking into the mantle. If the continents sank, the ocean would rush in to cover the land.

Figure 8-11 – Earth before mountains settle

Verse 25, discusses the above and provides scientific detail about the mountain forming process: "Before the mountains were settled in place, before the hills, I was given birth."[1]

The original Hebrew word for the English words "were settled in place" means to descend or fall. The word also has a connotation of to secure or to affix. The Hebrew word also implies covering or obscuring (as in drowning). This is a very apt description of what happened. Science tells us of a term called isostasy. This refers to landmasses settling into the Earth to provide sufficient support to support them. This process is like someone stepping into a boat. The boat sinks into the water to support the extra weight of the occupants. The same thing happened after the ocean upheaval initially pushed up mountains far above the Earth's surface. In the case of mountains, a very large part of them sank into the Earth. The original mountains probably appeared to drown. The Bible gives an apt description of what happened.

Figure 8-11 depicts what the newly formed jagged hills and mountains may have looked like.

---

[1] Proverbs 8:25

Figure 8-12 - Earth aftermountains settled

Figure 8-12 shows what Wisdom may have witnessed after the mountains settled into place.

Formation of Soil

We previously discussed how microbial mats, during the days of Vaalbara, first started breaking down the rocky surface of the Earth into dust. This first dust probably looked more like slime than soil. It was, however, the very beginnings of soil on Earth. The soil-making process accelerated with the decay of organic material from plants and animals, which lived and died to form more soil. As well, other forces of nature such as wind, running water, rain, and glaciers formed more and more soil. Proverbs speaks of a much higher quality than the soil produced by microbial mats: "I was given birth, before he made the world or its fields or any of the dust of the Earth."[1]

The original Hebrew word for "world" means the Earth in a geographical sense. This part of the verse takes us back to before the continents formed. The word for "fields" means a field or plot of agricultural land, separate from a town or city. It has a connotation of division by roads or fences. Wisdom here takes us forward in time to when fields formed. However, to produce food, the fields required a high caliber of soil.

---

[1] Proverbs 8:25–26

The phrase "or any of the dust of the Earth" is somewhat different in the original Hebrew Bible. The Hebrew word used for the English words "any of" means the "highest ranking" or "best part of." The Hebrew word for the English word "earth" means world, in the sense of being habitable, globe-friendly, and sustaining to life. A more apt translation for the phrase is "the best parts of the dust that makes up the land of the globe." The Bible speaks of the richest, highest quality soil that sustains us today. This soil resulted after millions of years of glaciation, water, and wind erosion and the decay of dead organic material forming soil.

How Does the Bible Fit So Well with Science?

The authors of the original Bible text lived before the pyramids. They did not have the insights from modern science that we do today. The medieval state of science at the time, could not possibly have given them the information to write what they did. The original Bible could reflect today's science, only if the hand of the LORD God guided their writings. The LORD God obviously had a master plan.

**Figure 8-13
Thales - an ancient philosopher**

King Solomon, the king of Israel, wrote Proverbs between 970 and 930 BC. The Iron Age had just begun. The Egyptians were starting to build the pyramids. The birth of Greek science started in about 600 B.C., nearly four hundred years after King Solomon wrote Proverbs. Figure 8-13[1] shows Greek philosopher Thales, traditionally regarded as the founder of Greek philosophy, was accomplishing his work. He speculated about the universe and assumed that it was possible to understand the universe using simple rules. Historians say that Thales proposed that the stars were other worlds.[2]

---

[1] https://en. wikipedia. org/wiki/Thales- Thales by Ernst Wallis et al is in the public domain because it was published 1875-9. The copyrights have expired.
[2] http://www. daviddarling. info/encyclopedia/T/Thales. html

Up until the time of Thales, people thought stars were simply lights suspended from a celestial vault. Thales explained earthquakes by hypothesizing that the Earth floats on water and those earthquakes occur when waves rock the Earth. Given the recent discovery of the deep, Thales may have been partially right. Thales did not know about the deep and so he probably viewed land as floating in the oceans.

Figure 8-14 Anaximander, astudent of Thales

In 580 B.C., Thales student Anaximander[1] postulated the idea that animal life sprang out of the sea long ago. He theorized that a spiny bark trapped the first animals before birth. As the animals grew older, the bark dried up and broke. As the early humidity subsided, dry land emerged and humankind had to adapt.

Anaximander introduced the notion that the Earth is not flat but cylindrical and that it floated free and unsupported at the exact center of the universe. He thought that people lived on one of the flat ends of the cylinder. He thought the sun was as large as the Earth.

In 500 B.C., Pythagoras postulated that the Earth is spherical, the perfect form. Heraclides[2] proposed that the rotation of the Earth on its axis caused the perception on Earth that the stars moved in the sky. Aristotle's theories did not agree. Aristotle postulated a fixed Earth as well as possible fixed stars and planets as well. Simplicius proposed that the Earth moves while the sun stays still.

Science, during the time of Proverbs and Psalms, was obviously at a primitive stage compared to today. Human knowledge of the time could not have possibly contemplated the origins of the universe as we know it today. Then how did King Solomon and King David have the insight to write what they did in Proverbs? The only plausible answer is that God inspired him to write the text, which remarkably fits with the actual formation of the universe, as science tells us.

---

[1] https://en. wikipedia. org/wiki/Anaximander- Anaximander by unknown author is in the public domain because it was published 1875-9. The copyrights have expired. – published in the U. S. before 1923 and public domain in the U. S.

[2] https://en. wikipedia. org/wiki/Heraclides_Ponticus

If the relatively few parts of the Bible that deal with the formation of the universe, the Earth and things in it are correct, then what does this say about the Bible? Perhaps the Bible is divinely inspired in its entirety!

The LORD's Master Plan

A long lasting and seemingly chaotic series of processes formed the universe and then the Earth. Although the processes forming the Earth seem random and haphazard, the result is far from random and haphazard.

Figure 8-15 shows a cross section of the Earth drawn to scale. The inner core is a solid. The outer core is liquid. Both the upper and lower mantle are solid. "The deep" I refer to in this book is a rocky material, inundated with water, otherwise known as the transition zone. The lower crust covers the entire globe. It is very thin under the oceans at about five to ten kilometers. The upper crust, which comprises all the landmasses on Earth does not show up in figure 8-15 because it is so thin. There is only a very thin film of upper crust scattered along the ocean floor.

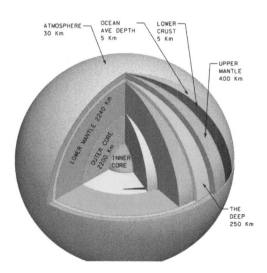

Figure8-15 Earth's layers

The upper crust, ocean, and atmosphere are amazingly thin compared to other inner layers of the Earth. Figure 8-15 easily distinguishes the thickness of the inner layers. The thicknesses of the outer layers are barely distinguishable in the scale image shown in figure 8-15. This gives an indication of how fragile our Earth's ecosystem is. Figure 8-16 shows the ocean at its average depth and the 30-kilometer-thick layer of the atmosphere containing most of the Earth's air.

To an observer on the ground, the atmosphere seems very high above the Earth's surface. We fly in jet planes far above the ground, and there is still enough oxygen to run the plane's engines. The ocean is so deep, that man cannot go to the deepest depths without special equipment, to withstand the severe pressure. Yet in the overall picture, the thickness of the atmosphere and depth of the oceans are insignificant in comparison to the size of the Earth. Indeed, the Earth is infinitesimal compared to the size of space.

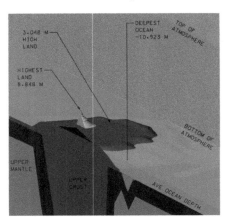

Figure 8-16 Earth's upper layers

Figure 8-16 shows the upper layers of the Earth, including land, which comprises the upper crust. The atmosphere now appears quite thick when viewed at a larger scale. Land when viewed to scale takes on a new perspective. The highest land elevation equivalent to the top of Mount Everest does not look so daunting. The Marianas Trench in the Pacific Ocean, although very deep, does not appear that significant when viewed from an overall perspective. The elevation of the flat portion of land shown is 3,048 meters above sea level. When viewed from an overall perspective, land seems to protrude just a small distance above the ocean. The average elevation of land on the globe is less than a thousand meters above sea level, significantly less that the land shown in figure 8-16. Even the land we live on is very thin and fragile in the overall scheme of things.

The creation of the universe, the subsequent formation of the Earth, and then the formation of things on the Earth follow a progression of events that seem to lead to a final goal. The LORD's initial creation of the universe by the Big Bang created a seemingly boundless universe. Out of this universe came the solar system and the Earth. Obviously, the hot gases that formed the universe and the Earth could not support any kind of life. Gradually the Earth cooled down, and things became more habitable for life. Eventually single-celled organisms appeared. These organisms caused the infertile rocky landscape of the Earth to start to break down into dust that would eventually result in soil. Single celled organisms evolved into multicelled organisms. Multicelled organisms and their successors evolved into plants and animals. The first single celled organisms that formed the soil made this possible. The final stages of evolution resulted in the modern animals we see today.

Then we see how modern man lives today. Life would not exist if not for the water deposited on Earth by asteroids and comets. We would not have land to live on if continents had not formed and became securely separate from the oceans. The plants we depend on for food would not exist if the simple-celled organisms of old did not break down the rocky terrain of the newly formed Earth into soil. None of the modern conveniences we have today would be possible without the minerals and metals we mine out of the Earth today. The strategic placement of iron ore and other minerals allowed man to make iron, steel, and other products. The same applies for a multitude of other metals and minerals such as bauxite to make aluminum, nickel, gold, silver, and a host of other metals and minerals. What about uranium, coal, oil, and natural gas? These products provide the energy to power vehicles, heat homes, and to make other products such as plastics. Humankind could not live the life it does in colder countries such as Canada were it not for the energy to heat our homes and businesses. None of these resources would exist if not deposited in the right spot at the right time.

The idea of a series of asteroid bombardments of the Earth seems like a chaotic and random occurrence. Yet all these events occurred at the right time and in the right order to deposit the right materials, where we as humans could use them billions of years later. Does this seem like the kind of event that occurred by chance? It seems unlikely to me. It seems more likely that the source or the LORD had a master plan. This the reason humankind lives as we do today?

As the Bible tells us, "By wisdom the LORD laid the Earth's foundations, by understanding he set the heavens in place; by his knowledge the watery depths were divided, and the clouds let drop the dew."[1]

---

[1] Proverbs 3:19

9

Conclusion

The Bible Is the Living Word

My work on this book convinced me more than ever that the Bible is the inspired Word of the Lord and God. As such it is indisputably accurate as written. The Bible, though, is the living Word. It keeps on speaking to you, the more you study it. Right to the very time I decided to publish this book, new insights prompted me to make changes to the book. I am sure that I would never publish this book, if I waited until I exhausted every possible detail the Bible could possibly reveal. Nevertheless, I feel that the general narrative of this book is correct. I strengthened my faith by writing this book. I am sure that as I continue to study the Bible, things will appear that further improve the details of this book. I also look forward to exploring other parts of the Bible to see what revelations might emerge. I am excited to see if a study of other parts of the Bible yields insightful results, such as in this book. Hopefully this will further the word of the LORD.

Proving that the Bible and science agree was the objective of this book. I feel more than ever that I succeeded in this goal. However, I am sure I have not exhausted every detail of this subject. I hope that this book is a starting point in a discussion. I hope this book inspires readers to delve into the Bible themselves. I am sure that if they do, the knowledge of the LORD will increase. This can only improve the future of mankind.

For Those Who Want to Explore More

The first manuscript for this book included much more detailed discussion about the origin and meaning of Hebrew words. Unfortunately, this cluttered the narrative, and often the details obscured the message. As a result, I removed much of the detail to make the message clear. Readers can explore the Hebrew wording of the Bible themselves by using various Bible interlinear on the Internet, such as www.biblehub.com. This allows access to *Strong's Exhaustive Concordance*, which provides definitions for Hebrew words.

Strong's Exhaustive Concordance, though, often lumps many Hebrew words under one definition. For example, the concordance uses only on definition for "heaven," but the original Hebrew Bible uses a number of different words for heaven. The first manuscript attempted to include the meanings of various individual words, where the concordance included only one definition.

THE REAL BIBLE AND SCIENCE

Thousands of Years Ago	God's Days	Persian Gulf	Civilizations	Settlements	Adam
3.80			Great Flood to Normal	End of Sumer	Noah/Descendants
4.00				Various Settlements	
4.20					
4.40				Kish, Uruk, Ur and	
4.60					
4.80				Great Flood	
5.00			Post Fld Dynasties	Jamdet Naser	
5.20				Larsa, Sippar	
5.40			Falling Levels		Seth to Noah
5.60	Seven			Uruk	
5.80					Badtibira
6.00					
6.20					
6.40					Adam Re-Born
6.60	Six				Eridu
6.80			Inland Fresh Water Flood		
7.00				Ubaid - South, Halif - North	
7.20					
7.40					Adam Formation to Spiritual Re-Birth
7.60					
7.80	Five				
8.00					
8.20					
8.40			Eden Flooded		
8.60	Four				
8.80					
9.00					
9.20			Pre-Pottery Neolithic		
9.40					
9.60				Mureybet - Neveli Cori	
9.80	Three				
10.00					
10.20					
10.40					
10.60					
10.80					
11.00	Two	Rising		Jerf el Amar	
11.20					
11.40					
11.60					
11.80					
12.00					
12.20	One		Natufian	Abu Hureyra	
12.40					
12.60					
12.80					
13.00					
13.20				None - Hunter Gatherer Peoples	
13.40					
13.60					
13.80		Dry			
14.00					

Citations

Armstrong, H. A., & D, B. M. (2005). *Microfossils.* Malden, MA, USA: Blackwell Publishing.

Barrientos, G. -M. (2012). *The Archaeology of Cosmic Impact: Lessons from Two Mid-Holocene Argentine Case Studies.* Springer Science+Business Media.

Beard, C. (2004). *The Hunt for the Dawn Monkey.* Berkeley and Los Angeles, California, USA: University of California Press.

Bobrowsky, D. P. (2007). *Comet and Asteroid Impacts and Human Society.* Ottawa/Uppsala, Ontario/Sweeden: Springer.

Bucaille, D. M. (n.d.). *The Bible The Qur'an and Science.*

Condie, K. C. (2011). *Earth as an Evolving Planetary System.* Oxford, UK: Academic Press.

de Vries, B. L. (2012). Comet-like mineralogy of olivine crystals in an.

Department of Ecology and Evolutionary Biology - University of California. (n.d.). *Ancient DNA.* (B. Shapiro, Ed.) Sanat Cruz, California, USA.

Eicher, D. J. (2013). *Comets - Visitors From Deeppace.* New York: Cambridge University Press.

Fasihnikoutalab, M. H. (2015). *New Insights into Potential Capacity of Olivine in Ground Improvement.* Department of Civil Engineering, University Putra Malaysia.

Harris, M. (n.d.). *The Nature of Creation.*

Hill, T. W. (1977). *Solar Wind Interactions.* Washington, DC, USA: National Academy of Science.

Hughlett, T. M. (2016). *Sensitivity of the Younger Dryas Climate to Changes in Freshwater, Orbital and Greenhouse Gas Forcing in Comprehensive Climate Models.* Arlington: The university of Texas Arlington.

Judaica Encyclopaedia. (2007). *Sumer, Sumerians.*
Kemp, T. (2005). *The Origin and Evolution of Mammals.* New York, NY, USA: Oxford University Press.
Palumbi, S. R. (2014). *The Extreme Life of the Sea.* Princeton, NJ, USA: Princeton University Press.
Springer. (2007). *The Black Sea Flood Question:.* Dordrecht, The Netherlands, The Netherlands: Springer.
Strong. (n.d.). *Strong's Exhaustive Concordance.*
Sundsdal, K. (2011). The Uruk Expansion: Culture Contact, Ideology and. *Norwegian Archaeological Review,* http://www.tandfonline.com/action/showCitFormats?doi=10.1080/00293652.2011.629812.
Swamy, K. S. (2010). *Physics of Comets.* (J. V. Narlikar, Ed.) Prune, India: World Scientific Publishing Co. Pte. Ltd.
Woolfson, M. M. (2013). *Time Space Stars and Man.* London, UK: Imperial College Press.

Index

Symbols

168 Lost Years 127

A

Abraham 111
Abu Hureyra 68, 69, 74
Acanthostega 147
Aira super volcanoes 44
Alulim 104
Anaximander 192
anomaly 44
Antarctica 41, 110, 126
anteater 153
anti-quark 172
aquatic tetrapods 147
Aratta 130
archaeothyris 149
Archosauriformes 151
ark 22, 110, 111, 112, 113, 114, 115, 117, 127
asteroid belt 183
Australopithecus 157, 158
Austriaco, Nicanor Pier Giorgio 19
Awan Dynasty 130

B

Babil Governorate 128
Badtibira 106
Barringer Ranch Crater 43

Big Bang 17, 22, 23, 30, 170, 171
binder 166, 167, 168
Black Sea 30, 61, 65, 95, 116, 117

C

Cambrian Explosion 144, 146
Cardinal Joseph Ratzinger 19, 20
Castorocauda 154
cement 71, 72
Chelyabinsk meteor 45
chimpanzees 155, 156, 157
Chogha Mish 90
Code of Ur-Nammu 133
coma 49, 50, 56
combine 168
continent phases 30
created 23, 24, 25, 26, 27, 29, 30, 41, 83, 93, 108, 109, 178
creationists 27
Cretaceous/Paleogene crisis 148
Cretaceous-Paleogene extinction 137, 153
cumulonimbus clouds 57
cuneiform 100
cynognathus 150

D

deep, the 192
dinosaurs 31, 76, 137, 148, 150, 151, 152

dirty thirties 51
Dryas 44, 68, 71

E

Eana 129
Egyptians 191
Einstein, Albert 171
Eirdu 87
electrons 171, 172
El Kowm 66
eukaryotic 142, 143
Euphrates River 67, 68, 69, 80
exosphere 56
exothermic reaction 72

F

Fertile Crescent 67
Flash Gordon 75, 76
fountains of the deep 121, 122, 181, 185
Francevillian biota 143

G

Garden of Eden 81
Gihon River 80
glacial cycles 38
Gobekli Tepe 70
gorillas 155, 157
grass family 60
great oxygenation event 141
Gregorian calendar 33, 34, 35
Grypania fungi 143

H

Haikouichthys 146
Halif period 87
Halif tournet 87

Halley's Comet 48, 50
Hassuna period 86
heavenly host 23, 24
Hebrew calendar 127
helium 172
herbs 60, 61
Hitchcock, Edward 115
homininae group 162
Homo erectus 159
Homo habilis 158
Homo heidelbergensis 160, 161
Homo sapiens 83
Huronian Ice Age 37
Hutton, James 32
hydrogen 172

I

ice age 35, 37, 39, 78
ice core studies 35
inner core 193
Inostrancevia 150
interglacial cycles. *See* glacial cycles
isostasy 189

J

jellyfish 145
Jemdet Nasr 109
Jerf el-Ahmar 68
Jubal 90
Jupiter 183
Jurassic Park 76
Jushur 128

K

Khafajah 128
King Enmebaragesi 128
King Meshkianggasher 129
King Solomon 21

L

Larag (Larsa) 106
lithium 172
lower crust 193

M

magic wand theory 27
magnetic field 55
magnetosheath 55
mammaliaforms 153
mammals 148, 153
mammary glands 152, 153
man of low degree 28, 107, 108
mantle 120, 183
McCormick, Cyrus 166
McDonald, Arthur B. 172
mechanical reaper 166
Mediterranean Sea 70, 116
mesons 172
Mesopotamia 65, 67, 71, 78, 81, 86, 87, 91, 92, 95, 100, 101, 103, 105, 106, 109, 110, 117, 127, 128, 129
Mesopotamian Bronze Age 90
mesosphere 56
microbial mats 140
Milky Way 172
molluscs 144
mountains 80, 114
Mount Ararat 110, 117, 127
Mount Everest 118, 119, 194
Mount Sinai 132
Mount St. Helens 42
multicelled organisms 195
Mureybet 69

N

Neanderthals 158, 160, 161
Neolithic Age (New Stone Age) 67
ners 103, 104
neutrinos 171, 172
neutrons 172
Nevalı Cori 70
New Testament 21
noctilucent cloud 56
Nod 86

O

Old Testament 21
Olivine 71, 72, 73
Opabinia 145
orders 154
Oruanui eruption 44
outer core 193
oxygenic photosynthesis 141
ozone layer 56

P

Paleolithic Age (Old Stone Age) 67
Permian crisis 148
Permian-Triassic extinction 137, 150
Persian Gulf 81, 94, 106
Pishon River 80
Plesiadapis 155
Pottery Neolithic 68
President Trump 133
priestly story 21
primate order 155
prokaryotic cells 141, 142
protons 172
Pythagoras 192

Q

quark 172
Quaternary extinction event 41
Quaternary Ice Age 37
Quaternary mass extinction 41

R

Rio Cuarto Comet 41
RNA 138, 139
Ross Ice Shelf 126

S

Sacardotal Bible 112
Sahelanthropus 156
Samarra period 87
sars 128
Saturn 183
sauropsid group (reptiles) 148
scythe 164, 166
sea ice 126
Seth 108
sexually reproduction 141, 142
sheaves 167
Shuruppak 106, 113
sickle 69, 70, 164
Simplicius 192
Snowball Earth 137
solar system 48, 49, 175, 177, 178, 179, 181, 183, 195
solar winds 55
springs of the deep 118, 121, 123, 185, 186
Sputnik 76
starry host 23, 173, 174
stooks 167
Strait of Hormuz 78
Sudbury Neutrino Observatory 172
Sumer 100, 101, 102, 103, 106, 128, 133
Sumerian culture 89, 100, 103
Sumerian King List 103, 128, 129
super critical point 119
synapsids. *See* mammals

T

Tanakh 20, 21
Taupo Volcano 44
Temple 1 comet 49
terrestrial planets 183
tetrapods 147
Thales 191
therapsids 150
thousand-year to a day concept 34
Tigris River 80
time beyond knowledge 171
Toba Volcano 41, 43, 44
Tower of Babel 129
transition zone 120
Tubal-Cain 90
Turkey 70, 117, 127

U

Ubaid culture 92
upper crust 181, 193, 194
Ur-Schatt River 80, 81
Uruk 99, 106, 107, 129
Urukagina 133

V

Vaalbara 141, 184, 187, 190

W

Wadi Al-Rummah River 80
Westothiana 148
wisdom 176

Y

Yahvist Bible 112

Z

Zimbir 106
Ziusudra 113, 115

About the Author

Richard grew up on a mixed farm in Saskatchewan, Canada, where he gained an appreciation of nature and learned about hard physical work. An upbringing in a staunch Catholic environment provided a basic understanding about religion. Richard attended grades 1 to 5 in a one-room country schoolhouse before completing grades 6 to 12 in a more modern, one-class-per-room school, at Annaheim, Saskatchewan. An engineering degree at the University of Saskatchewan provided the basis for a career in the Canadian oil industry.

Richard always questioned things, including religion. At the age of forty or so, he left the Catholic church and joined the Worldwide Church of God, a smaller non-denominational church. There he gained a deeper understanding of the Bible as opposed to the religious ritual and dogma-based orientation of the Catholic Church. The Worldwide Church of God broke up, and Richard became an ever more committed non-church attending Christian.

Richard's interest in Christianity and learning about the Bible grew after he stopped participating in organized religion. Approximately four or five years ago, he started working on a book about the Bible and science. A long career as an engineer and project manager in the oil and gas industry provided strong critical thinking and organizing skills, which allowed him to analyze many scientific disciplines and compare them to the Bible. This resulted in the book *The Real Bible and Science*.

Richard enjoys discussing the book with people. Those who read the manuscript were impressed with the contents and encouraged him to proceed with publishing the book. Richard enjoys conversations about the Bible and Christianity and always takes the opportunity to discuss these subjects with others.